HEALTH CAREERS CHEMISTRY

An Introduction to General, Organic, and Biological Chemistry

LAB MANUAL

Roman E. Boldyreff | Assistant Professor of Biology
Cuyahoga Community College

Title page image: Dr. William N. Boldyreff, center stage, the author's grandfather, lectures at Tokyo University February 21, 1921. Seated to his left is Ephraim B. Boldyreff, the author's father.

Health Careers Chemistry

An Introduction to General, Organic, and Biological Chemistry
Cuyahoga Community College
Roman E. Boldyreff

Printed in the United States of America
10 9 8 7 6 5 4 3 2 1
ISBN: 978-1-61740-828-1

Van-Griner Learning
Cincinnati, Ohio
www.van-griner.com

President: Dreìs Van Landuyt
Project Manager: Janelle Lange
Customer Care Lead: Lauren Houseworth

Boldyreff 828-1 Su19
314209
Copyright © 2020

BIO1100
Table of Contents

Name: _______________________________________

crn: _______________________________________

Outcomes
Apply problem solving skills to succeed in preparatory coursework for the healthcare professions or other scientific studies.

Experiments, like many everyday activities, involve some element of risk. Driving a car, preparing supper, or mowing the lawn all involves some degree of risk. We minimize risk by being aware of potential dangers, knowing and following safety precautions, and being fully aware and ready to use emergency procedures when necessary. One of the most important actions you can take to ensure your safety and the safety of others is to read over each laboratory exercise before you start any assignment. Your instructor will alert you to safety considerations for each experiment you perform in this class and inform you of laboratory safety precautions and emergency procedures, but it is up to you to learn and follow them.

1. Eating, drinking, or smoking while doing an experiment can distract you and others. To insure a safe laboratory environment, never bring food or drinks into the laboratory and never smoke in the lab. This includes electronic vaporizing devices.

2. Dressing for a lab is also important. Shoes must be worn at all times in the lab. Sandals or opened-toed shoes are not permitted nor is loose clothing. Long hair and neck ties should be tied up and secured.

3. Your instructor will tell you when to wear eye protection. Eye protection in a chemistry lab consists of chemical resistant, splash proof safety goggles. Safety glasses are not goggles and over only limited protection for your eyes.

4. Fume hoods are designed to contain noxious and toxic odors and experimental mishaps. Your instructor will point out the location of the fume hoods. Make sure any reactions involving dangerous chemicals or unpleasant odors are performed in the fume hood.

5. Many common reagents, for example, alcohols, acetone, and especially ethers, are highly flammable. Do not use them near open flames. When using a Bunsen burner always pick it by the base. Never hold the burner by the tip.

6. Only carry out the experiments you are instructed to perform and only enter your scheduled lab when your instructor is present. No student should enter the stock or prep rooms.

7. Always replace the cap or lid on any container securely each time you use it. Consider all chemicals to be hazardous unless instructed otherwise. Avoid cross contamination of solutions.

8. Balances are sensitive, keep the area around the balances clear and clean. Never place chemicals or any specimen directly on the balance pan. Always use a plastic mass dish or clean glassware.

9.   All chemical spills, must be cleaned up immediately. If the chemical involved is in any way toxic, consult your instructor.

10.   There is no breakage fee for glassware. If you break glassware, report it to your instructor and you will be given instructions on how it should be cleaned up.

11.   Thoroughly familiarize yourself with the location and operation of the safety equipment in your lab.

12.   If chemicals come in contact with your skin or eyes, wash immediately with large amounts of water and then notify your instructor.

13.   If you get an irritating substance in your eyes, wash it out with the emergency eyewash. The emergency shower is used, if large areas of your body are exposed to caustic chemicals.

14.   The fire extinguisher is used to put out equipment fires in the lab. The emergency gas cut off button is used in the case of a gas fire. Electric outlets are ground fault interrupter type and will break the circuit in case of an overload.

15.   Always clean up your lab space and return all materials and supplies, but not test solutions and used reagents, to the place from where you obtained them.

16.   Some test solutions and reagents may be poured down the sink but many are toxic. Your laboratory instructions will inform you when you are working with toxic or hazardous substances. If you have any doubt if a substance is toxic, ask your instructor. Dispose of all toxic materials in the appropriate marked container located in the fume hood.

17.   Glassware should be rinsed and returned to the tray marked "USED GLASSWARE."

18.   Microscope slides and cover slips used for making wet mounts are disposable. Dispose of slides and cover slips in the container marked "USED SLIDES". Prepared microscope slides are to be returned to their proper tray. Microscopes should be placed in the standard storage position, returned to their cabinet, and stored with a cover.

19.   Always wash your hands and clean up your lab space after each lab session.

20.   If an accident occurs in the lab, tell your lab instructor no matter how trivial you think it may be. In the case of a major accident, your instructor may ask you to call Campus Safety. You should learn the phone number for your Campus Safety Office.

I have read the above and I will adhere to the rules and procedures described.

Signature: ____   **Date:** ________________________

Printed Name: ____

CRN: _________________________________   **Lab Day & Time:** _______________________________

BIO1100
Lab Equipment

Name: ____

crn: ____

<table>
<tr><td align="center">Outcomes</td></tr>
<tr><td>Apply problem solving skills to succeed in preparatory coursework for the healthcare professions or other scientific studies.</td></tr>
</table>

Successfully completing the lab exercises in this course will require the use of a variety of lab ware. The College provides students with this equipment and assembles it into trays. The following equipment can be found in your assigned tray:

Beakers	50, 100, and 250 ml. Beakers are used for several purposes- to contain substances, for mixing, and for heating or cooling but you should always remember that the measurements on beakers are approximate.
Erlenmeyer flasks	125 and 250 ml. Erlenmeyer flasks are conically shaped with flat bottoms. The markings on Erlenmeyer flasks are also approximate.
Graduated cylinders	10, 25, 50, and 100 ml. Graduated cylinders are used for measuring volume. You should be able to measure to the 0.1 ml when using them.
Tongs	Tongs may be used to pick up Erlenmeyer flasks and small beakers but should never be used to pick up test tubes.
Test tubes	Twelve 16 x 120mm. An approximate measurement on test tubes this size is that one finger width is about 2-3 ml.
Test tube holder	Used to pick up test tubes. Always aim the open end of a test tube away from yourself and others when heating it. Never use a test tube holder to pick up beakers or flasks.
Test tube brush	Used to clean your test tubes. Clean your test tubes with glass ware powder not hand soap. Rinse test tubes thoroughly with deionized/distilled water. Do not dry the inside of your test tubes with paper towels.
Spatula	Your spatula is used to transfer small amounts of substances from their storage containers to test tubes, beakers, and flasks. It should not be used to stir mixture because many mixtures can corrode metals.
Glass Rod	Mixtures are stirred with a glass stirring rod. A stirring rod can also be used as aid to decant liquids without spilling them.
Funnel	Used as an aid when pouring and decanting liquids and powders.
Drying Crystal	Used to separate solutes from solvents. Also referred to as a watch glass

In addition, to your individually assigned items, laboratory equipment and supplies for specific labs is located on auxiliary laboratory benches and tables. If you look at the back and side benches you will see test tube racks, hotplates, dry-water baths, beaker holders, thermometers, and assorted rings and clamps. Remember to return items and supplies promptly after you are finished using them so that others may use them. At your personal bench you will find strikers and Bunsen burners.

Printed Name: ____

Tray #: _________

3

Notes:

BIO1100
Lab 1 Measurement

Name: _________________________________

crn: _________________________________

Outcomes

Apply problem solving skills to succeed in preparatory coursework for the healthcare professions or other scientific studies.

A. Prelab Assignment

The drawing below represents a section of a graduated cylinder. Each mark on the cylinder is one ml. The longer line between 20 and 30 ml marks 25 ml. First notice the curved nature of the water line. The curved water line is named the meniscus. Meniscus is Latin for curve. The formation of a meniscus is a result of the cohesive and adhesive properties of water. The most accurate measurement of the volume of the water in a graduated cylinder is to read the volume at the bottom of the meniscus. Also notice that the graduate is read at eye level.

What is the volume of water in the graduated cylinder? There is clearly more than 25 ml in the graduate so to record the volume as 25 ml would be inaccurate. A more accurate description of the volume would be to estimate how much the meniscus is over the 25 ml mark. If the meniscus was exactly half way between the 25 ml mark and the 26 ml mark, the volume could be read as 25.5 ml. If the meniscus was exactly on the 25ml mark, the volume would be read as 25.0 ml. Would you say the volume was 25.2 ml or 25.3 ml? Either reading would be acceptable but 25.1 ml or 25.4 ml would be inaccurate. Notice that a measurement has both a number and a unit.

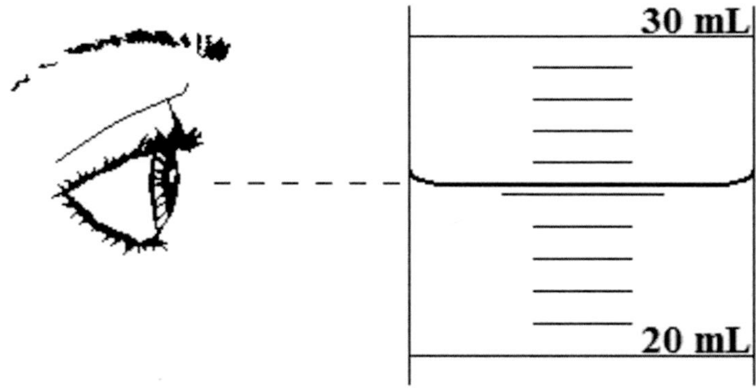

Record the volume of each graduate cylinder, the first one has been done for you

1.

Volume = _____ **23.7 ml**

(Note that the answer includes the unit.)

2.

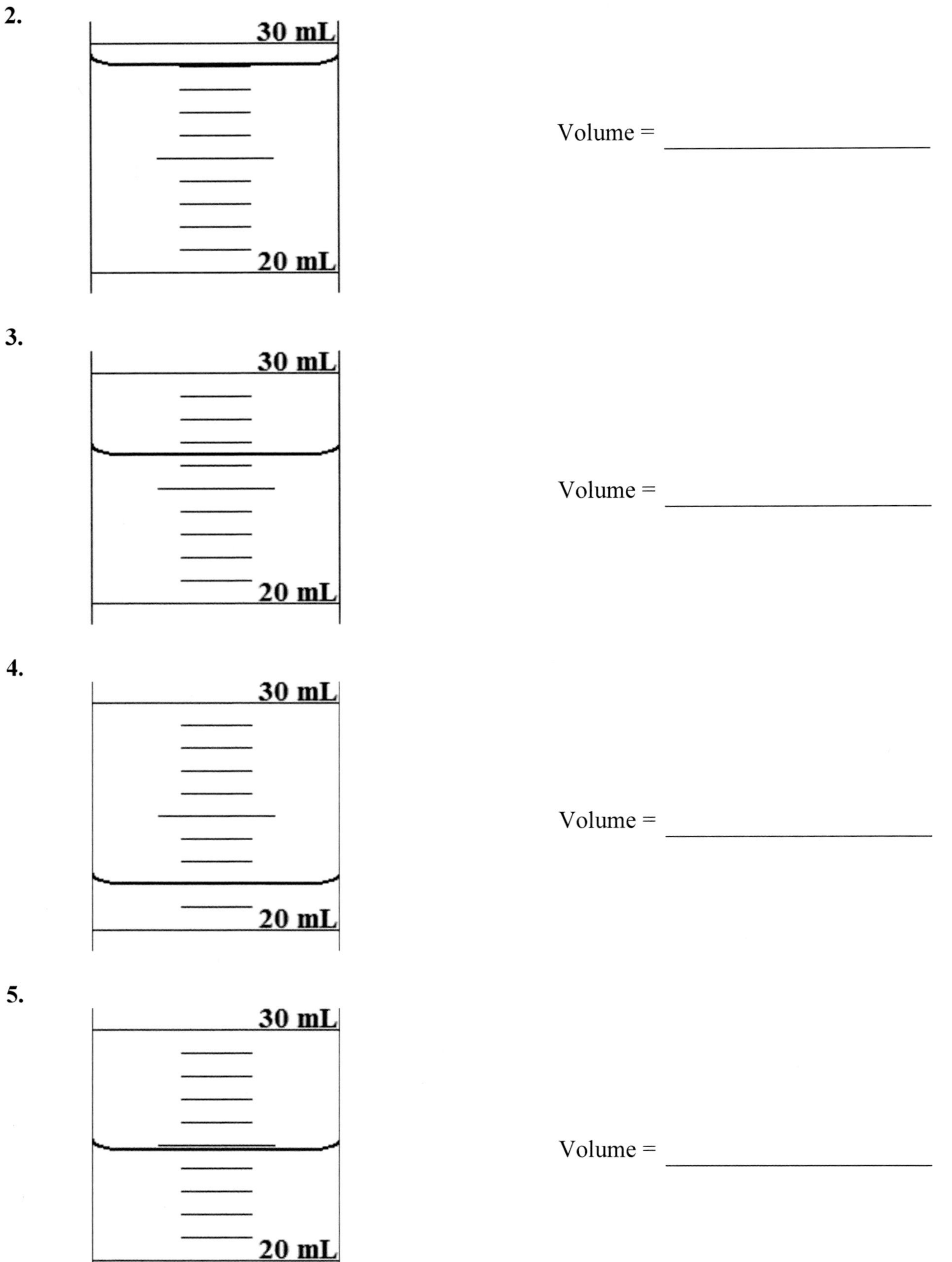

Volume = _______________________

3.

Volume = _______________________

4.

Volume = _______________________

5.

Volume = _______________________

    The same method used to read graduated cylinders is used to read glass thermometers. Record the temperature of each thermometer to the nearest one-tenth of a °C(0.1 °C) and include the temperature unit in your answer.

6.   **7.**

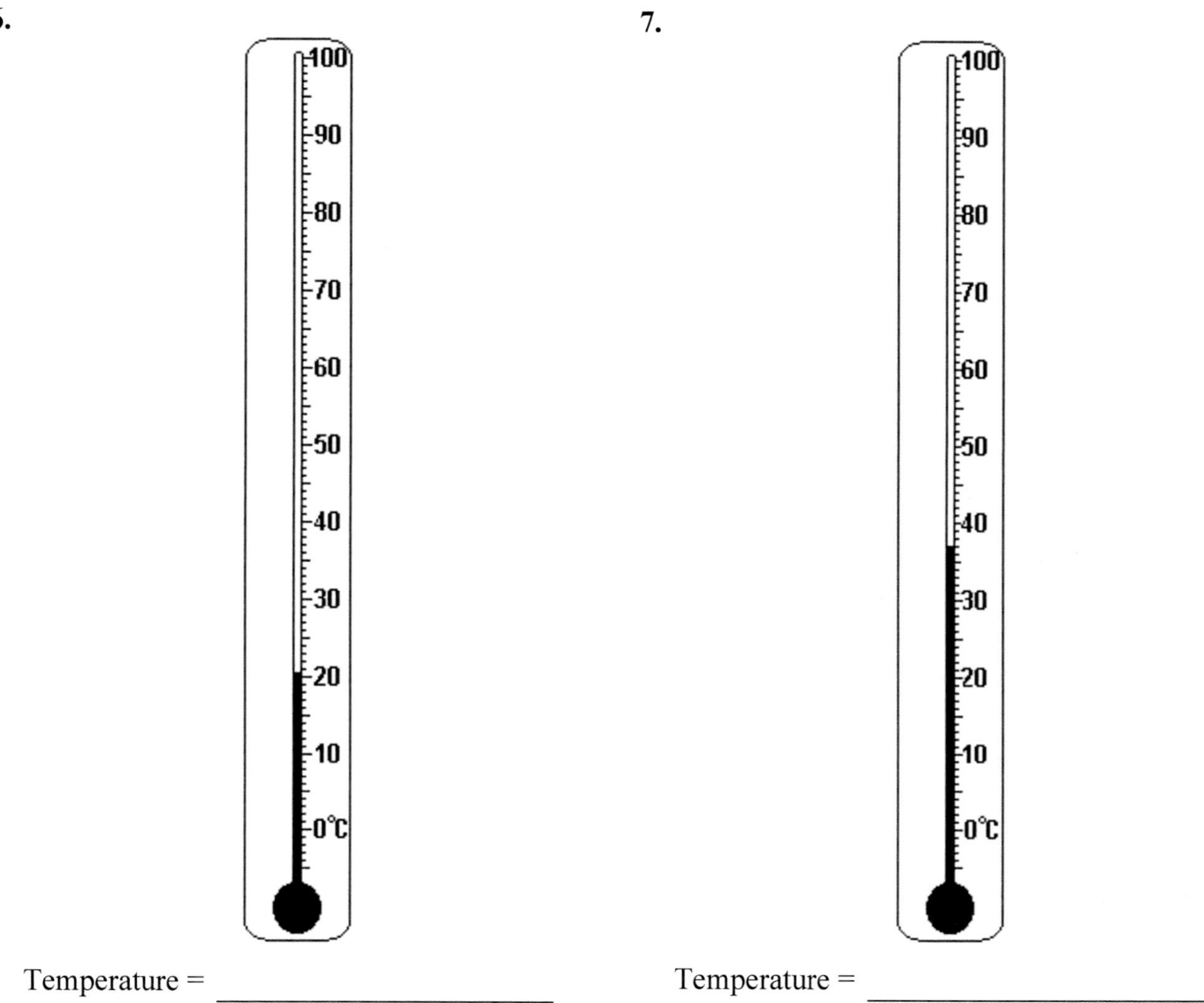

Temperature = _______________   Temperature = _______________

B. Measuring Volume

1. Beaker versus Graduated Cylinder- Fill a 250 ml beaker with water to the 50 ml mark, transfer the water from the beaker into a 100 ml graduated cylinder. Next you are going to read the volume of the transferred water that is in the graduated cylinder. When reading the volume in your graduated cylinder always position your head such that your eye is level with the bottom of the rounded, concave liquid surface, called the bottom of the meniscus. Failure to do this simple procedure will result in an inaccurate reading.   Read the volume to the closest 0.1 ml, and record your answer below.   Remember, a measurement includes a number and the unit of measure.

Volume = _______________

2. Flask versus Graduated Cylinder- Fill a 250 ml Erlenmeyer flask with water to the 50 ml mark, transfer the water from the flask into a 100 ml graduated cylinder. Next you are going to read the volume of the transferred water that is in the graduated cylinder. Remember to position the graduated cylinder at eye level. Read the volume to the closest 0.1ml, and record your answer below.

Volume = _______________

3. Accuracy- The closeness of a measurement to the actual quantity being measured is the measurement's accuracy. The closer a measurement is to the quantity being measured, the more accurately the measurement describes that quantity. Of the beaker, flask, and graduated cylinder, which device gave the most accurate description of a volume of fifty milliliters of water? Record your answer on the line below.

C. Measuring Density

Density is a ratio of the mass of a quantity to its volume. Mathematically, stated:

$$\text{Density} \quad = \quad \frac{\text{Mass}}{\text{Volume}} \quad = \quad \frac{g}{ml}$$

1. Obtain a rubber stopper from the lab bench.

2. Measure and record the mass of the rubber stopper using one of the balances located in the lab. The reading should be to the closest one hundredth of a gram (0.01 g). Remember to include the unit of measurement in you answer.

Mass of rubber stopper = ______________

3. Carefully fill a 100 ml graduated cylinder exactly to the 50.0 ml mark. Remember to read the graduated cylinder at eye level. Record the volume of the water in the 100 ml graduated cylinder. Remember that a measurement includes a number and a unit.

Volume of water = ______________

4. Hold the graduated cylinder at a shallow angle and carefully slide the rubber stopper into the water. Be careful not to lose any water. Set the graduated cylinder on a level surface, read, and record the new volume to the nearest 0.1 ml.

Volume of water & rubber stopper = ______________

5. Subtract the volume of water in **3.** from the volume of water & rubber stopper in **4.** Record the difference of the two volumes below. Your answer should be to the 0.1 ml and include the unit of measurement.

Volume of rubber stopper = ______________

6. Divide the mass of the rubber stopper in **2.** by the volume of the rubber stopper in **5.** to calculate the object's density. Record the density below. Remember that a calculated value cannot have more significant digits than the value with the least significant digits used in the calculations. The units of density are g/ml.

Density of rubber stopper = ________________

7. Return your lab tray to its cabinet.

C. Temperature Conversion

You may be familiar with the Fahrenheit to Celsius temperature conversion formula shown below:

$$F = 9/5\ C + 32$$

Could you use the temperature conversion formula above to convert Celsius to Fahrenheit? Carefully read the instructions below and then complete the temperature conversion given below.

Fahrenheit to Celsius- 1. Add 40 to the °F temperature
2. Multiply the sum by 5 °C/9 °F
3. Subtract 40 from the product

Celsius to Fahrenheit- 1. Add 40 to the °C temperature
2. Multiply the sum by 9 °F /5 °C
3. Subtract 40 from the product

Example: Given 98.6 °F
Fahrenheit to Celsius-

1. Add 40 to the °F temperature	98.6 °F + 40 = 138.6 °F
2. Multiply the sum by 5 °C/9 °F	138.6 °F (5 °C/9 °F) = 77.0 °C
3. Subtract 40 from the product	77.0 °C - 40 = <u>37.0 °C</u>

Example: Given 22 °C
Celsius to Fahrenheit-

1. Add 40 to the °C temperature	22 °C + 40 = 62 °C
2. Multiply the sum by 9 °F /5 °C	62 °C (9 °F /5 °C) = 111.6 °F
3. Subtract 40 from the product	111.6 °F - 40 = <u>72 °F</u>

Complete the following temperature conversions using the above method. Show your calculations to receive credit. The line under the last zero indicates that it is significant, so 10<u>0</u> has three significant digits. A line under the second zero, 1<u>0</u>0, would be two significant digits. While 100 would have one significant digit.

1. 10<u>0</u> °C   = ________________ °F (Water boils)

2. 32 °F   = ________________ °C (Water freezes)

3. 72 °F   = ________________ °C (Room temperature)

4. -40 °C   = ________________ °F (Surprised?)

D. Metric and Conventional Conversion

Table of Conversion Factors

Conventional	=	Metric
39.37 in	=	1.000 m
1.00 in	=	2.54 cm
1.0567 qt	=	1.000 l
1.000 fluid oz	=	29.57 ml
2.205 lb	=	1.000 kg
1.0000 lb	=	453.59 g
1.000 oz	=	28.35 g
0.03527 oz	=	1.000 g

The factor conversion method is conversion technique that uses cancellation of units. Study the examples below and then complete the conversions listed. Show your calculations to receive credit.

Example 1: Convert 72 ½ inches to centimeters.

72.5 in   x $\dfrac{2.54 \text{ cm}}{1.00 \text{ in}}$ =   184 cm

Example 2: Convert 175 pounds to kilograms.

175 lb   x $\dfrac{1.000 \text{ kg}}{2.205 \text{ lb}}$ =   79.4 kg

1. 25_0 ml   = ________________ fl oz

2. 31.3 kg   = ________________ lb

3. 9_0 g   = ________________ oz

4. 15.0 cm   = ________________ in

5. 12 in   = ________________ m

6. 12_000 g   = ________________ lb

Notes:

Name: _______________________________

crn: _______________________________

Outcomes

Recognize and name ionic compounds.
Properly combine ions to form a balanced ionic formula.
Write and balance a chemical equation using appropriate symbols and recognize major types
  of chemical reactions.

A. Prelab Assignment

Use printed or internet sources to define the terms below.

1. Atom: ____

____

____

____

2. Element: ____

____

____

____

3. Molecule: ____

____

____

____

4. Compound: ____

____

____

____

5. Ion: ____

____

____

____

B. Naming Ionic Compounds

Some molecular formulas are complicated like the molecular formula of turquoise: $CuAl_6(PO_4)_4(OH)_8$, but most are simple. To write the formula of an ionic compound, first consider the charges of the positive ion and the negative ion in the compound. Next, remember that the resulting molecule must be electronically neutral. That is, the positive charge must cancel the negative charge. Last, write the symbol for the positive ion first without showing its charge and use a subscript, if needed. Following the positive ion; write the negative ion's symbol without its charge and a subscript if needed. To name the compound, use the name of the positive ion but for the negative ion, use the –ide suffix. For compounds containing polyatomic ions, simply name the ion. A list of common ions with their charges and names is located at the end of this lab. Study the following examples, and then fill in the tables below the examples.

Example 1- What would be the formula for a compound formed by sodium and chlorine and what would it be named?

First- Sodium has a charge of a positive one and is written as its elemental symbol with a superscript positive sign, Na^{+1}. The ion of chlorine is chloride and it has a negative one charge and is symbolized, Cl^{-1}.

Next- Since the sodium ion has a positive one charge and the chloride ion has a negative one charge, only one of each is needed to make the resulting molecule electronically neutral.

Last- The symbol for the positive member of the compound is written first followed by the symbol for the negative member and the compound is named-

NaCl, sodium chloride

Example 2- What would be the formula for a compound formed by potassium and sulfur and what would it be named?

First- Potassium has a charge of a positive one and is written as its elemental symbol with a superscript positive sign, K^{+1}. Sulfur's ion is named sulfide and its symbol , S^{-2}.

Next- Since the sodium ion has a positive one charge and the sulfide ion has a negative two charge, it will take two K^{+1} ions to cancel the S^{-2} ion.

Last- The symbol for the positive member of the compound is written first, with a subscript of two, followed by the symbol for the negative member and the compound is named-

K₂S, potassium sulfide

Example 3- What would be the formula for a compound formed by calcium and phosphorus and what would it be named?

First- Calcium has a positive two charge and is written as its elemental symbol with a superscript positive two, Ca^{+2}. The ion of phosphorus is named phosphide, has a negative three charge, and is symbolized as P^{-3}.

Next- There are two methods used to figure out what subscripts are needed in a situation like this. One method is to multiply the value of the charge superscripts: 2 x 3 = 6. This tells you that you will need 6 units of positive charge and 6 units of negative charge to make the resulting molecule electronically neutral. Think, what times +2 will give a +6 and what times -3 will give a -6. The answers will tell you what subscripts you need. An alternative method is to drop the charges of the superscripts, switch them, and write them as subscripts. Either technique, will give you the subscripts required.

Last- The symbol for the positive member of the compound is written first, with a subscript of three, followed by the symbol for the negative member, with a subscript of two, and the compound is named-

Ca_3P_2, calcium phosphide

Example 4- What would be the formula for a compound formed by barium and nitrate ion and what would it be named?

First- Barium has a positive two charge and is written as its elemental symbol with a superscript positive two, Ba^{+2}. The ion polyatomic nitrate ion is composed of three oxygen atoms covalently bonded to a single nitrogen atom. It has a negative one charge and is written as NO_3^{-1}.

Next- Since the barium ion has a positive two charge and the nitrate ion has a negative one charge, it will take two NO_3^{-1} ions to cancel the Ba^{+2} ion.

Last- The symbol for the positive barium ion is written first, followed by the symbol for the negative nitrate ion which is enclosed in parenthesis, and a subscript of two, and the compound is named-

$Ba(NO_3)_2$, barium nitrate

Example 5- What would be the formula for a compound formed by ammonium and sulfur and what would it be named?

First- Ammonium has a positive one charge and is written as NH_4^{+}. The sulfur anion has a negative two charge and is written as S^{-2}.

Next- Since the ammonium ion has a positive one charge and the sulfide ion has a negative two charge, it will take two NH_4^{+1} ions to cancel the negative S^{-2} ion.

Last- The symbol for the positive member of the compound is written first, enclosed in parenthesis, with a subscript of two, and followed by the symbol for the negative member and the compound is named-

$(NH_4)_2\ S$, ammonium sulfide

Exercise 1- Using the given symbols and charges for the ions below, write the formulae and names for the ionic compounds that would result from combining the ions.

	Br^{-1}	S^{-2}	P^{-3}
Na^{+1}	Formula:		
	Name:		

	Cl^{-1}	O^{-2}	$C_2H_3O_2^{-1}$
Ca^{+2}	Formula:		
	Name:		

	MnO_4^{-1}	CO_3^{-2}	PO_4^{-3}
H^{+1}	Formula:		
	Name:		

	Cl^{-1}	O^{-2}	$C_2H_3O_2^{-1}$
Fe^{+2}	Formula:		
	Name:		

Exercise 2- Using the given symbols and charges for the ions below, write the formulae and names for the ionic compounds that would result from combining the ions.

		Cl^{-1}	SO_4^{-2}	PO_4^{-3}
K^{+1}	Formula:			
	Name:			

		Br^{-1}	NO_3^{-1}	$C_2H_3O_2^{-1}$
Mg^{+2}	Formula:			
	Name:			

		OH^{-1}	PO_4^{-3}	HCO_3^{-1}
NH_4^{+1}	Formula:			
	Name:			

		OH^{-1}	CO_3^{-2}	CN^{-1}
H^{+1}	Formula:			
	Name:			

C. Naming Covalent Compounds

Several different compounds can be formed by the covalent bonding of the atoms of just two different elements.   The same two elements can form compounds having different numbers of atoms of each different element so the naming of covalent compounds uses prefixes to indicate the relative number of each element's atoms. Consider the combinations of nitrogen and oxygen that result in compounds.

Molecular Formula	Name
N_2O	Dinitrogen monoxide (Nitrous oxide)
N_2O_3	Dinitrogen trioxide
N_2O_4	Dinitrogen tetraoxide
N_2O_5	Dinitrogen pentaoxide
NO	Mononitrogen monoxide (Nitric oxide)
NO_2	Mononitrogen dioxide

Example 1- What would be the molecular formula and name for a compound formed by a single carbon atom and two sulfur atoms and what would it be named?

First-   Carbon is to the left of sulfur on the periodic table and is less electronegative than sulfur. Carbon has three oxidation-reduction (redox) states (+4, +2, and -4). Since carbon in its -4 state cannot combine with sulfur, we do not consider the -4 redox state. So carbon must be in a +2 or +4 state. If carbon is in +2 redox state, it would combine with only one sulfur atom. Since a single atom of carbon has combined with two sulfur atoms, the carbon atom must be in a +4 redox state.

Next-   Sulfur is to the right of carbon on the periodic table and is more electronegative than carbon. Sulfur has four redox states (-2, +2, +4, and +6). Since two sulfur atoms are combining with one +4 carbon atom, the sulfur atoms must be in a -2 redox state.

Last-   The symbol for the less electronegative carbon atom is written first, followed by the symbol for the more electronegative sulfur atoms and a subscript of two.

Molecular Formula: CS_2 **Name:**   carbon disulfide

Example 2- What would be the formula for a compound formed by a single sulfur atom and two oxygen atoms and what would it be named?

First-   Sulfur is below oxygen in the periodic table and is less electronegative than oxygen. Sulfur has four redox states (-2, +2, +4, and +6). Since sulfur in its -2 state cannot combine with oxygen, we do not consider the -2 redox state. So sulfur must be in a +2, +4, or +6 state. If sulfur is in +2 redox state, it would combine with only one oxygen atom and its +6 redox state it could combine three oxygen atoms.   Since a single atom of sulfur has combined with two oxygen atoms, the sulfur atom must be in a +4 redox state.

Next-   Oxygen is above sulfur in the periodic table and is more electronegative than sulfur. Oxygen has four redox states (-1, -2, +1, and +2). Since oxygen is only in a +1 or +2 when combined with the Group VIIA halides and since oxygen is only in a -1 redox state as a peroxide when it combines with Group IA or IIA atoms to from peroxides, such as hydrogen peroxide, H_2O_2, the oxygen atoms must be in a -2 redox state.

Last-   The symbol for the less electronegative sulfur atom is written first, followed by the symbol for the more electronegative oxygen atoms and a subscript of two, and the compound is named-

Molecular Formula: SO_2 **Name:**   **sulfur dioxide**

Example 3- What would be the formula for a compound formed by two nitrogen atoms and three oxygen atoms and what would it be named?

First-   Oxygen is to the right nitrogen in the periodic table so it must be in one of its negative redox states, either -1 or -2. Remember, oxygen is only in a -1 redox state as a peroxide when it combines with Group IA or IIA atoms to from peroxides so the oxygen atoms must be in its -2 redox state and we can consider the three oxygen atoms to have a total electronegativity of -6.

Next-   Nitrogen is to the left oxygen in the periodic table and is less electronegative than oxygen. Nitrogen has eight redox states (-1, -2, -3, +1, +2, +3, +4, and +5). Since nitrogen must be in one of its positive redox states, it must be in a +1, +2, +3, +4, or +5 state. Since the three oxygen atoms have a total of -6 electronegativity, the two nitrogen atoms must have a +3 charge each for a total of +6.

Last-   The symbol for the less electronegative nitrogen atoms are written first with a subscript of two, followed by the symbol for the more electronegative oxygen atoms and a subscript of three, and the compound is named-

Molecular Formula: N_2O_3 **Name:**   **dinitrogen trioxide**

Exercise 1- Write the formulae of the following compounds:

1.   Dinitrogen monoxide (nitrous oxide) _______________________________________

2.   Diboron trioxide   _______________________________________

3.   Boron arsenide   _______________________________________

4.   Mononitrogen dioxide   _______________________________________

5.   Sulfur dioxide   _______________________________________

Exercise 2- Write the names of the following compounds:

1. SiO_2 _______________________________________

2. CO _______________________________________

3. CS_2 _______________________________________

4. NO _______________________________________

5. CO_2 _______________________________________

D. Balancing Equations

Chemical reactions involve the separation, rearrangement, or recombination of atoms. Metabolism refers to all the chemical reactions that take place in an organism. In a chemical reaction, new substances, products, are formed from old substances, reactants. Chemical equations are a shorthand method to illustrate a chemical reaction. The reactants are written to the left of an arrow and the products are written to the right of the arrow.

$$\text{Reactants} \longrightarrow \text{Products}$$

In a balanced chemical equation, the reactants and the products are represented by elemental symbols or molecular formulae. The elemental symbol or the molecular formula represent exactly one mole of the reactant or product. Coefficients are used to adjust the molar ratios of the reactants and products so that the relative amount of each reactant or product has the same amount of matter (the same number and type of each atom) in the reactants as in the products. Molar coefficients are always written in the lowest whole number ratio. For example, 6 to 4 would be reduced to 3 to 2.

Steps to balancing chemical equations:	Do not:
1. Name the reactants and products	1. Try to balance all of the atoms at once
2. Write valid molecular formulas	
3. Adjust molar coefficients	2. Change the compound formulas when adjusting molar coefficients

Example 1- Iron ore is purified to pure iron by reducing iron III oxide (Fe_2O_3) with carbon (C) to produce iron (Fe) and the by-product carbon monoxide (CO).

Step 1- Write a reaction equation by simply naming the reactants and the products. Examine the example below and match the terms to the written description above.

$$\text{Iron III oxide} \quad + \quad \text{Carbon} \longrightarrow \text{Iron} \quad + \quad \text{Carbon monoxide}$$

Step 2- Use your chemistry knowledge or a reference text to help replace the compound names with their appropriate **molecular or compound formulas.** It is absolutely critical that these formulas are accurate for the compounds involved in the reaction.

$$Fe_2O_3 \quad + \quad C \quad \longrightarrow \quad Fe \quad + \quad CO$$

Step 3- Use whole number **molar coefficients** to adjust the relative amount of each reactant or product to ensure we have the same amount of matter (the same number and type of each atom) in the reactants as in the products. Always start at the left and go to the right. On the reactant side there are two Fe, so a 2 is written in front of the Fe on the product side.

$$Fe_2O_3 \quad + \quad C \quad \longrightarrow \quad 2\,Fe \quad + \quad CO$$

Starting from the left again. The Fe symbols are balanced. Moving to the left, there are three O in Fe_2O_3 on the reactant side, so a 3 is written in front of CO on the product side.

$$Fe_2O_3 \quad + \quad C \quad \longrightarrow \quad 2\,Fe \quad + \quad 3\,CO$$

Starting from the left again, the Fe symbols are balanced as are the O symbols but there is one C on the reactant side and three C on the product side. Placing a 3 in front of the C on the reactant side balances the equation. The equation can be checked by starting at the left and moving to the right.

$$Fe_2O_3 \quad + \quad 3\,C \quad \longrightarrow \quad 2\,Fe \quad + \quad 3\,CO$$

Example 2- Sodium nitrate ($NaNO_3$) can be produced from the reaction of calcium nitrate ($Ca(NO_3)_2$) with sodium chloride ($NaCl$). The by-product is calcium chloride ($CaCl_2$).

Step 1- Write a reaction equation by simply naming the reactants and the products. Examine the example below and match the terms to the written description above.

$$\text{Calcium nitrate} \quad + \quad \text{Sodium chloride} \quad \longrightarrow \quad \text{Sodium nitrate} \quad + \quad \text{Calcium chloride}$$

Step 2- Use your chemistry knowledge or a reference text to help replace the compound names with their appropriate **molecular or compound formulas.** It is absolutely critical that these formulas are accurate for the compounds involved in the reaction.

$$Ca(NO_3)_2 \quad + \quad NaCl \quad \longrightarrow \quad NaNO_3 \quad + \quad CaCl_2$$

Step 3- Use whole number **molar coefficients** to adjust the relative amount of each reactant or product to ensure we have the same amount of matter (the same number and type of each atom) in the reactants as in the products. Starting at the left and going to the right. On the reactant side there is one Ca and there is one Ca on the product side, so nothing needs to written since the molecular formula by itself is understood to be one mole.

$$Ca(NO_3)_2 \quad + \quad NaCl \quad \longrightarrow \quad NaNO_3 \quad + \quad CaCl_2$$

Starting from the left again. The Ca symbols are balanced. Moving to the left, there are two NO_3 in $Ca(NO_3)_2$ on the reactant side, so a 2 is written in front of $NaNO_3$ on the product side. Note that the NO_3 was balanced as a single polyatomic ion. The N and O were not balanced separately.

$$Ca(NO_3)_2 \quad + \quad NaCl \quad \longrightarrow \quad 2\,NaNO_3 \quad + \quad CaCl_2$$

Starting from the left again, the Ca symbols are balanced as are the NO_3 symbols but there is one Cl on the reactant side and two Cl on the product side. Placing a 2 in front of the NaCl on the reactant side balances the equation. The equation can be checked by starting at the left and moving to the right.

$$Ca(NO_3)_2 \quad + \quad 2\,NaCl \quad \longrightarrow \quad 2\,NaNO_3 \quad + \quad CaCl_2$$

Exercise 1- Balance the following equations:

1. $H_2 \quad + \quad O_2 \quad \longrightarrow \quad H_2O$

2. $Zn \quad + \quad H_2SO_4 \quad \longrightarrow \quad ZnSO_4 \quad + \quad H_2$

3. $KClO_3 \quad \longrightarrow \quad KCl \quad + \quad O_2$

4. $Mg \quad + \quad O_2 \quad \longrightarrow \quad MgO$

5. $Mg \quad + \quad HCl \quad \longrightarrow \quad MgCl_2 \quad + \quad H_2$

6. $F_2 \quad + \quad H_2O \quad \longrightarrow \quad HF \quad + \quad O_2$

7. $NH_3 \quad + \quad NaOCl \quad \longrightarrow \quad NaOH \quad + \quad NCl_3$

8. $H_2S \quad + \quad O_2 \quad \longrightarrow \quad SO_4^{-2} \quad + \quad H_2O$

10. $CH_4 \quad + \quad O_2 \quad \longrightarrow \quad CH_3OH \quad + \quad CO_2 \quad + \quad H_2O$

Exercise 2- Balance the following equations:

1. $HC_2H_3O_2$ + H_2SO_4 $\longrightarrow$ CO_2 + H_2S + H_2O

2. CH_4 + NH_3 + O_2 $\longrightarrow$ HCN + H_2O

3. H_2SO_4 + $Al(OH)_3$ + KOH $\longrightarrow$ $AlK(SO_4)_2$ + H_2O

4. NH_3 + $NaOCl$ $\longrightarrow$ $NaOH$ + NH_2Cl

5. Al + O_2 + H_2O $\longrightarrow$ $Al(OH)_3$

6. LiC + $LiCoO_2$ $\longrightarrow$ Li_2O + CoO + C

7. H_2S + CO_2 $\longrightarrow$ $C_6H_{12}O_6$ + S + H_2O

8. CH_4 + O_2 $\longrightarrow$ CO_2 + H_2O

9. $HC_3H_5O_3$ $\longrightarrow$ $HC_3H_5O_2$ + $HC_2H_3O_2$ + CO_2 + H_2O

F. Appendix 1 – Common Ions

Cations (Positive Ions)		Anions (Negative Ions)	
Ion	**Symbol & Charge**	**Ion**	**Symbol & Charge**
Aluminum	Al^{+3}	Acetate	$C_2H_3O_2^{-1}$
Ammonium	NH_4^{+1}	Bicarbonate	HCO_3^{-1}
Barium	Ba^{+2}	Borate	BO_3^{-3}
Calcium	Ca^{+2}	Bromide	Br^{-1}
Chromium (II) or Chromous	Cr^{+2}	Carbonate	CO_3^{-2}
Chromium (III) or Chromic	Cr^{+3}	Chlorate	ClO_3^{-1}
Copper (I) or Cuprous	Cu^{+1}	Chloride	Cl^{-1}
Copper (II) or Cupric	Cu^{+2}	Citrate	$C_6H_5O_7^{-3}$
Hydrogen	H^{+1}	Cyanide	CN^{-1}
Iron (II) or Ferrous	Fe^{+2}	Fluoride	F^{-1}
Iron (III) or Ferric	Fe^{+3}	Hydroxide	OH^{-1}
Lead (II) or Plumbous	Pb^{+2}	Iodide	I^{-1}
Lead (IV) or Plumbic	Pb^{+4}	Lactate	$C_3H_5O_3^{-1}$
Lithium	Li^{+1}	Nitrate	NO_3^{-1}
Magnesium	Mg^{+2}	Nitrite	NO_2^{-1}
Manganese	Mn^{+2}	Oxide	O^{-2}
Mercury (I) or Mercurous	Hg^{+1}	Perchlorate	ClO_4^{-1}
Mercury (II) or Mercuric	Hg^{+2}	Permanganate	MnO_4^{-1}
Potassium	K^{+1}	Phosphate	PO_4^{-3}
Silver	Ag^{+1}	Phosphide	P^{-3}
Sodium	Na^{+1}	Pyruvate	$C_3H_3O_3^{-1}$
Tin (II) or Stannous	Sn^{+2}	Sulfate	SO_4^{-2}
Tin (IV) or Stannic	Sn^{+4}	Sulfide	S^{-2}
Zinc	Zn^{+2}	Sulfite	SO_3^{-2}

G. Appendix 2 – Redox States of Common Nonmetals

Nonmetal	Symbol	Redox states
Arsenic	As	**-3**, **+3**, or, +5
Boron	B	**+3**
Bromine	Br	**-1**, **+1**, **+3**, +5, or +7
Carbon	C	**-4**, +2, or **+4**
Chlorine	Cl	**-1**, **+1**, +3, +5, or +7
Fluorine	F	**-1**
Hydrogen	H	-1 or **+1**
Iodine	I	**-1, +1,** +5, or +7
Nitrogen	N	**-3**, +2, **+3**, +4, or +5
Oxygen	O	-1 or **-2**
Phosphorus	P	**-3**, **+3**, or +5
Selenium	Se	-2, **+4**, or +6
Silicon	Si	**-4**, +2, or **+4**
Sulfur	S	-2, +2, **+4**, or **+6**

(Most common state in bold)

Notes:

Name: ____

crn: ____

> **Outcomes**
> Describe the similarities and differences between the states of matter.
> Write and balance a chemical equation using appropriate symbols and recognize major types
>   of chemical reactions.
> Discuss the role of energy in chemical reactions.

A. Prelab Assignment

Use printed or internet sources to define the terms below.

1. Endergonic Reaction: ____

____

____

____

2. Exergonic Reaction: ____

____

____

____

3. Enthalpy: ____

____

____

____

4. Entropy: ____

____

____

____

5. Gibbs Energy: ____

____

____

____

!!! SAFETY GOGGLES MUST BE WORN DURING THIS LAB!!!

B. Combination Reaction

In a combination reaction two or more reactants combine to form a single product or a major product and a minor product. The combustion of paper is an example of a combination reaction. At 233 $^{\circ}$C (451 $^{\circ}$F), there is sufficient energy for oxygen to directly combine with the cellulose in paper and turn the paper into ash releasing CO_2. The oxidation of antimony with chlorine gas is another example of a combination reaction.

$$Sb \quad + \quad 3\ Cl_2 \quad \longrightarrow \quad 2\ SbCl_3$$

At high temperature, O_2 will combine with Mg to produce MgO. Magnesium and oxygen are the reactants and magnesium oxide is the product. This reaction is both a combination reaction and a combustion reaction. Combustion is defined as any reaction that involves the combination of a substance with oxygen with the rapid release of light and heat. All combustion reactions are combination reactions but not all combination reactions are combustion reactions. This reaction can also be described as an exergonic or exothermic reaction since there is a net release of energy.

1. You should be wearing your safety goggles. Obtain a strip of magnesium metal from the lab bench. Carefully examine the magnesium ribbon and describe its appearance. What color is the magnesium? Would you say silver or gray? Describe the magnesium's luster. Is it dull or shiny? Is the magnesium rigid or pliable? Rigid means that it returns to its original shape when bent. Pliable is the opposite of rigid. Write your description on the lines below.

____

____

____

2. Light your Bunsen burn and adjust it until you have a small blue flame. The flame should be composed of a darker blue flame with a lighter blue flame within it. There should not be any orange in your flame. Adjust the air vent at the base of the burner until you have the flame just described. This will result in the high temperature required to activate this reaction.

!!! DO NOT LOOK DIRECTLY AT THE Mg FLAME !!!

3. Place your drying crystal on your lab surface close to your Bunsen burner. Use your tongs to clamp and hold the smallest portion of the Mg strip possible. Hold the end of the Mg strip between the tips of the light blue and dark blue flame. This part of the flame will give you the temperature required to activate the combination reaction of Mg and O_2 and produce a flame. Note the color of the flame but do not stare at the flame. The flame is evidence that a chemical reaction has occurred. Describe the flame.

____

____

____

4. When the Mg-O_2 flame has stopped, carefully place the product of the reaction in your drying crystal. Turn off your Bunsen burner. Note the appearance of the product, MgO. What color is MgO? The product is no longer a metal. How would you describe the composition of MgO? Write your description of the color and composition of MgO on the lines below.

____

____

____

5. Write the molecular formula for the product of the composition reaction you have just observed and balance the chemical equation.

$$Mg \quad + \quad O_2 \quad \longrightarrow$$

6. In the above reaction, atoms of oxygen in the diatomic O_2 molecule combined with atoms of Mg to produce a new compound. The reactants oxygen, a gas, and Mg, a metal, combine to form a white powder, MgO. Write the definition of a combination reaction below.

____

____

____

C. Decomposition Reaction

In a decomposition reaction a single reactant is decomposed or broken down to form two or more products. Potassium chlorate, $KClO_3$, a compound, will decompose to form potassium chloride, KCl, a compound, and oxygen, O_2, an element. In another decomposition reaction, lead II hydroxide, $Pb(OH)_2$,a compound, is decomposed into lead II oxide, PbO, a compound, and water, H_2O, a compound. In an electrolysis apparatus at sufficient voltage, water, H_2O, a compound, can be decomposed into its constituent elements, hydrogen gas, H_2, and oxygen gas, O_2.

The electron arrangement of hydrogen and oxygen in water give it a characteristic structure. The two hydrogen atoms are bonded to the oxygen at a $107.5°$ angle to each other. In the diagram below, the water molecules are drawn with oxygen and hydrogen atoms oriented as they would be in an electrical field. Without an electrical field, the water molecules would be turned in all different kinds of directions.

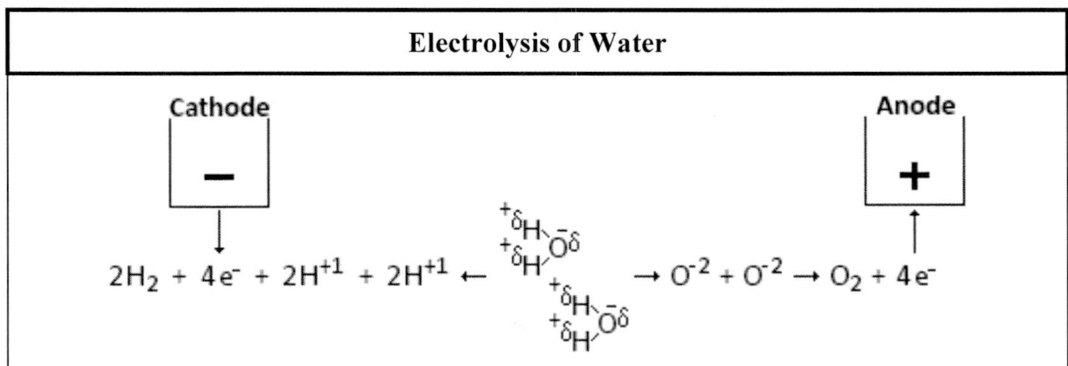

The arrangement of electrons in H_2O places the two hydrogen atoms in a plane below the oxygen atom and with a two pairs of lone electrons above the plane of the oxygen atom. This arrangement makes the water molecule very polar with a positive region around the two hydrogen atoms and a negative region around the two pairs of lone electrons of O_2. In an electric field with a strong potential difference (voltage), water molecules are pulled apart as the negative region is pulled towards the anode and the positive region is pulled towards the cathode. This forms cations (positively charged particles) and anions (negatively charged particles) from the water molecule.

In an electrolysis apparatus, cations are attracted to the negatively charged electrode or cathode and anions are attracted to the positively charged electrode or anode. At the cathode, hydrogen ions receive electrons and are reduced to form diatomic hydrogen gas. While, at the anode, oxide ions lose electrons and are oxidized to form diatomic oxygen gas. Your instructor is now going to demonstrate the electrolysis of H_2O.

1. You should be wearing your safety goggles. What is the volume of the gas collected from the cathode column? What is the volume of the gas collected from the anode column?

Volume of cathode gas, mL	Volume of anode gas, mL

2. According to the explanation given above, the gas collected at the cathode is H_2. While O_2 supports combustion, it does not burn, while H_2 readily burns if exposed to a flame. What evidence did you observe that the gas collected at the cathode was H_2? What evidence did you observe that the gas collected at the anode was O_2?

____

____

____

____

____

____

3. Gay-Lussac demonstrated that the reactants and products of elements combine in whole number ratios by volume. Avogadro reasoned that these volumes would have an equal number of particles. If the above statements are true then the ratio of the two volumes would be the same as the ratio of hydrogen atoms to oxygen atoms in a water molecule.

Volume of H_2 gas, ml	Volume of O_2 gas, ml
Ratio of Hydrogen:Oxygen =	

4. Write the molecular formula for the products of the decomposition reaction you have just observed and balance the chemical equation. Remember that both hydrogen and oxygen are diatomic.

$$H_2O \longrightarrow \qquad\qquad +$$

5. In the above reaction, water was decomposed to its constituent elements of hydrogen and oxygen. A single reactant water, a liquid, is separated to form two products, hydrogen, a gas, and oxygen, a gas. Write the definition of a decomposition reaction below.

____

____

____

D. Single Replacement Reaction

In a single replacement reaction, also called a single displacement reaction, one element replaces another element in a compound. The reaction of water, a compound, with calcium, an element, results in the production of calcium hydroxide, a compound. In this reaction, Atoms of the element Ca replace one of the two H atoms in H_2O. A common single replacement reaction is to react an acid and metal to produce a salt and hydrogen gas.

1. You should be wearing your goggles. Place a test tube (16 x 120 mm) in a test tube rack. Use a disposable pipette to carefully add 5 ml (about two finger widths) of 1.0 M HCl into the test tube. Use caution, HCl is a strong acid. Set the test tube back in the test tube rack. Place the test tube and rack in a safe place on your lab bench.

2. Light your Bunsen burn and adjust it until you have a small blue flame as before. Set the Bunsen burner in a safe place on your lab space. Obtain a strip of magnesium metal from the lab bench and carry it to your lab space.

3. Pick up the test tube containing HCl. Grip the test tube in your fist with your thumb covering the opening. Lift your thumb and place the Mg ribbon in the test tube. As soon as you add the Mg to the test tube, place your thumb back over the opening. Observe for approximately 30 seconds and consider the following questions- Is the test tube getting warmer? Is the Mg ribbon decreasing in size? Is there any activity in the test tube? Do you feel any pressure against your thumb? All of these are evidence that a chemical reaction is taking place. Write the evidence that you observed below.

____

____

____

____

____

____

____

4. Keep your thumb over the opening of the test tube containing the reacting Mg and HCl. Keep a tight seal on the opening for about 1 minute. You should still observe bubbles being generated in the solution in the tube. Now, you are going to see if the gas produced in this reaction is flammable. Make sure you are not pointing your test tube at anyone including yourself. Bring the top of the test tube towards the flame and remove your thumb as it gets close to the flame. What do you see? What do you hear? Dispose of the contents of the test tube in the laboratory bench sink drain. Record your observations below.

____

____

____

5. Write the molecular formula for the product of the composition reaction you have just observed and balance the chemical equation.

$$Mg \quad + \quad HCl \quad \longrightarrow$$

6. In the above reaction, atoms of the element magnesium, a metal, replace hydrogen in hydrochloric acid, a compound, to form magnesium chloride, a salt, and hydrogen, a diatomic gas. Write the definition of a single replacement (displacement) reaction below.

____

____

____

E. Double Replacement Reaction

In a double replacement reaction or double displacement reaction, the oxidized (positive) member of one of the reactant compounds replaces the oxidized (positive) member of the other reactant compound to produce two new product compounds. In the reaction of sodium chloride (NaCl) with silver nitrate ($AgNO_3$), the Na^+ in sodium chloride replaces the Ag^+ in silver nitrate. The products of this reaction are silver chloride (AgCl) and sodium nitrate ($AgNO_3$).   The sodium nitrate is soluble in water at room temperature but the silver chloride is not and precipitates or falls out of solution. You are going to investigate the reaction of sodium hydroxide and copper II sulfate.

1. You should be wearing your goggles. Place a test tube in a test tube rack. Use a disposable pipette to carefully transfer 4.0 ml of 1.0 M NaOH into a 10.0 ml graduated cylinder. Use caution, NaOH is a strong base. Carefully transfer the NaOH solution to the test tube (16 x 120 mm) in the test tube rack. Set the test tube rack in a safe place on your lab space. Place a thermometer in the test tube. Allow the thermometer at least a minute to equilibrate to the temperature of the NaOH solution. Read the temperature of the solution to the nearest 0.1 °C and record the temperature below. Remember a measurement has a number and a unit.

Initial Temperature = ________________

2. Rinse out the graduated cylinder with water and shake out any excess water into the sink. Use a disposable pipette to carefully transfer 2.0 ml of 1.0 M $CuSO_4$ into the 10.0 mL graduated cylinder. Add the 2.0 ml of 1.0 M $CuSO_4$ to the test tube with the NaOH. Stir the mixture gently but thoroughly with your thermometer. Read the temperature of the solution to the nearest 0.1 $^\circ$C and record the temperature below. Has the temperature of the solution increased or decreased? An increase would indicate an exergonic reaction while decrease would indicate an endergonic reaction.

Final Temperature = __________

3. Let the mixture in the test tube set for a few minutes in the test tube rack. Do you notice a precipitate settling to the bottom of the test tube? Answer this question after completing questions **4.** and **5.** below.

____

____

____

____

4. Write the molecular formula for the product of the composition reaction you have just observed and balance the chemical equation.

NaOH   + $CuSO_4$ $\longrightarrow$

5. In the above reaction, the Na^+ of sodium hydroxide replaced the Cu^{+2} of copper II sulfate. The products are copper II hydroxide and sodium sulfate. Write the definition of a double replacement (displacement) reaction below. The products of this double replacement reaction can be disposed of in the sink.

____

____

____

____

____

6. Return your lab tray to its cabinet.

F. Chemical Reactions & Energy

From your Pre-lab Assignment, you learned that all chemical reactions are either exergonic (exothermic) or endergonic (endothermic). You have just observed or carried out four reactions. After each type of reaction listed below, write if the reaction was exergonic or endergonic.

Reaction	Exergonic or Endergonic
Combination	
Decomposition	
Single Replacement	
Double Replacement	

Notes:

Name: _______________________________

crn: _______________________________

Outcomes
Apply the concept of tonicity to osmosis and dialysis.

A. Prelab Assignment

The square knot is a widely used knot that firmly. The knot is formed by tying a left-handed overhand knot and then a right-handed overhand knot, or vice versa. Study the figure below and then tie two pieces of string together using a square knot. You may wish to view an on-line tutorial on how to tie a square knot if you are having trouble.

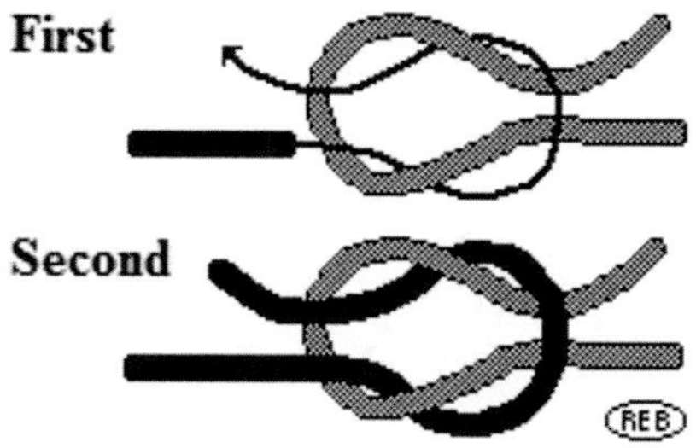

When you have successfully tied a square knot, save the joined strings, and bring them to lab. When you turn in your assignment, staple your square knot to the upper left hand corner of your assignment.

1. Square Knot _________________

GOGGLES MUST BE WORN DURING THIS LAB!!!

B. Dialysis

Diffusion is the net, random movement of the particles of a substance from an area of high concentration of that substance to an area low concentration of that substance. The diffusion of oxygen across the alveolar cells of the lungs and into the blood plasma and then into the hemoglobin of the erythrocytes is important physiologically. As is the diffusion of carbon dioxide out of body cells into blood plasma and out of blood plasma across the alveolar cell membranes and into the lungs.

Dialysis is the diffusion of small molecules, ions, and water through a differentially permeable membrane. Differential permeability is determined by particle size.

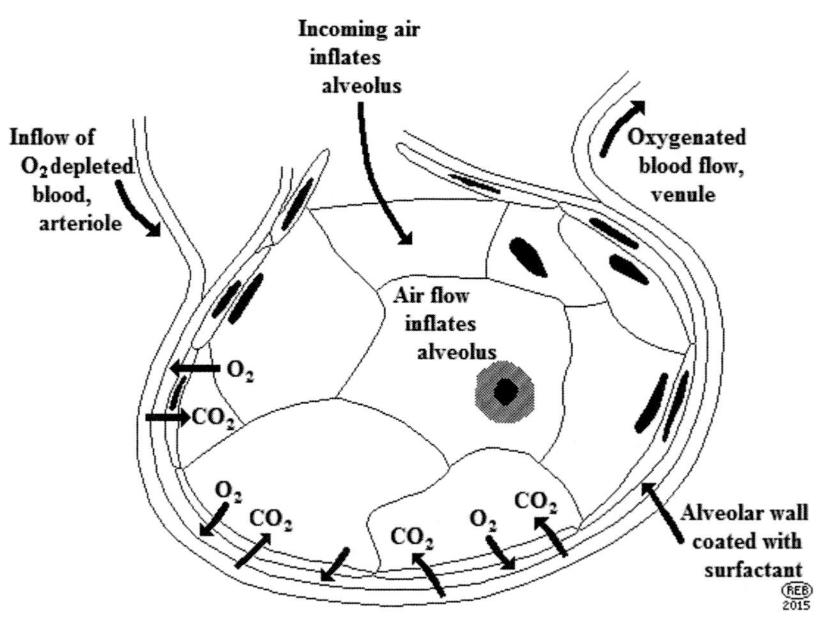

1. You should be wearing your safety goggles. Pair up with another student. Obtain a 10-12 cm strip of dialysis tubing and two 10-15 cm lengths of string from the supply bench. The dialysis tubing and string may already be cut to the desired length. If dialysis tubing and string are not already cut, use the scissors provided to cut them into the appropriate lengths. Carry dialysis tubing and string back to your lab station.

2. Take the dialysis tubing to your lab bench sink. Turn the water faucet on to a low flow of water. Hold the dialysis tubing between your thumb and fingers of your dominant hand. Place your hand down into the sink and let the water run over the dialysis tubing. Keep your hand low in the sink to avoid splashing water. Gently rub the tubing with your thumb and fingers. You should begin to feel the tubing loosen up and feel slippery. Continue to rub the dialysis tubing until it completely opens up.

3. Gently twist one end of the tubing to form about one cm of twisted tubing. Fold the twisted section of the tubing over on itself. Grasp the end of the folded tubing and the folded over end of the tubing between the thumb and fingers of both your hands. A small section of twisted, folded over tubing should be visible between your thumbs.

4. Have your lab partner tie a square knot around the twisted, folded over tubing that is between your thumbs. You should now have a piece of dialysis tubing that is closed on one end. Place the small end of your funnel into the open end of the dialysis tubing.

5. Return to the supply bench with your partner. Use a disposable pipette and carefully transfer two full droppers of NaCl into the dialysis tubing. Next, use a fresh pipette and carefully transfer two full droppers of glucose into the dialysis tubing. Lastly, gently use a fresh pipette and carefully transfer two full droppers of starch into the dialysis tubing.

6. Twist the open end of the dialysis tubing, fold the twisted end back on itself, hold the twisted end between your thumbs and fingers, and have your partner tie off the end with a square knot. You now have a sealed dialysis bag with a mixture of Na^+, Cl^-, glucose, and starch particles. Place the dialysis tubing in a 250 ml beaker. Add quantity sufficient RO water to cover the dialysis tubing. Do not use tap water. Let the tied-off dialysis tubing with the three mixtures set in the RO water for about 30 minutes. Continue to **C. Osmosis**, unless instructed otherwise

7. At the end of 30 minutes, remove the sealed dialysis tubing from the beaker and discard the dialysis tubing. Do not discard the water in the beaker. You are now going to test the water in the beaker to see which particles were able to pass through the differentially permeable membrane.

8. Place three test tubes in a test tube rack. Use a grease pencil and mark one tube with an "A" for $AgNO_3$ test, one tube with a "B" for Benedict's test, and one tube with an "L" for Lugol's test. Decant about 5 ml of the beaker water into each of the three tubes.

9. Add 8-10 drops of Benedict's test solution to the test tube marked with a "B." Note the blue color of the Benedict's test solution. Place this test tube into the dry water bath for about 20 minutes. Benedict's solution tests for the presence of simple sugars. Glucose is a monosaccharide simple sugar. The primary ingredient in Benedict's solution is copper(II)sulfate, $CuSO_4$. Many sugars are able to reduce (add an electron to) Cu^{+2} ions. The addition of the electron changes the charge on the Cu ion from +2 to +1 and the Cu^{+1} ions combine with dissolved oxygen in the water to form Cu_2O. Since the sugar lost an electron, it said to be oxidized. The aldehyde group on the sugar is changed to a carboxylic acid group. The reaction is shown below for glucose.

glucose + copper(II)sulfate   gluconic acid + copper(I)oxide

Since the Cu ions are reduced, any sugar that can do this is called a reducing sugar. Most but not all simple sugars are reducing sugars. Copper(II)sulfate solutions are blue and water soluble, while copper(I)oxide is reddish brown and insoluble in water. The insoluble Cu_2O molecules settle to the bottom of the test tube and form a red-orange precipitate. **A positive Benedict's test can range from a slight greenish blue to a heavy brick red.** Copper(II)sulfate will also oxidize aldehydes, so a positive Benedict's test can also indicate the presence of aldehydes. Record the result of your Benedict's test in the data collection table below in **12**.

10. Add 2-4 drops of $AgNO_3$ test solution to the test tube marked with a "A." Place this test tube back into the test tube rack. The $AgNO_3$ tests for the presence of Cl^- ion and is based on the following double replacement reaction. Record your $AgNO_3$ test result in the data collection table below in **12**.

$$AgNO_3 \quad + \quad NaCl \quad \longrightarrow \quad AgCl \quad + \quad NaNO_3$$

While the $NaNO_3$ product is soluble in water, the AgCl is not and forms a precipitate. **The formation of the AgCl precipitate causes the test solution to turn cloudy gray to white depending on the amount of AgCl formed.** A positive result indicates Cl^- ions have dialyzed out of the dialysis tubing and into the beaker water. Chloride ions are larger than sodium ions, so if you get a positive test for Cl^- particles you can assume that the Na^+ particles diffused though the membrane into the beaker water. Since NaCl dissociates in water to Na^+ and Cl^- particles, it is incorrect to say NaCl particles dialyzed through the membrane.

11. Add 2-4 drops of Lugol's solution (Iodine reagent) to the test tube marked with a "L." Lugol's solution ranges from yellow-orange to red-orange in color. Place this test tube back into the test tube rack. Lugol's solution tests for the presence of starch particles. Lugol's solution is prepared by mixing equal molar amounts of KI salt with elemental I_2. This mixture exists in a dynamic equilibrium of K^+ and I_3^- particles as shown below.

$$I_2 \quad + \quad KI \quad \rightleftharpoons \quad I_3^- \quad + \quad K^+$$

The triiodide anions complex with starch particles and turn the test mixture blue-black. **So a color change from orange to blue-black indicates that starch particles have dialyzed out of the dialysis tubing.** Record the result of your Lugol's test in data collection table below in **12**.

12. Record the results of your $AgNO_3$, Benedict's, and IKI tests in the data table below.

Test	Color	+ or -
$AgNO_3$		
Benedict's		
Lugol's		

13. Remember, dialysis is the diffusion of small molecules, ions, and water through a differentially permeable membrane. Differentially permeable means that the membrane allows some particles to pass through but stops others. Particles that are too large for the pores of the membrane cannot dialyze through the membrane. What particles dialyzed through the dialysis bag membrane?

____

14. What particles did not dialyze through the dialysis bag membrane?

____

15. What factor stopped particles from diffusing across the membrane?

____

16. Return your lab tray to its cabinet.

C. Osmosis

While dialysis is the diffusion of solute particles and solvent through a differentially permeable membrane, osmosis is the diffusion of solvent only through a differentially permeable membrane. The cellular environment can be described as isotonic, hypotonic, or hypertonic. Study the diagrams below and notice the differences in the three situations.

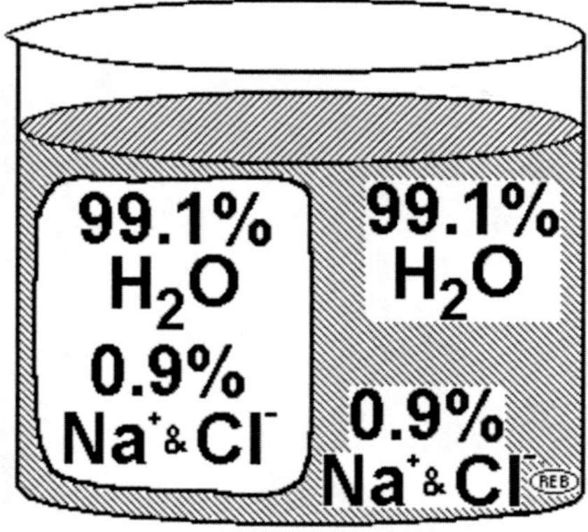

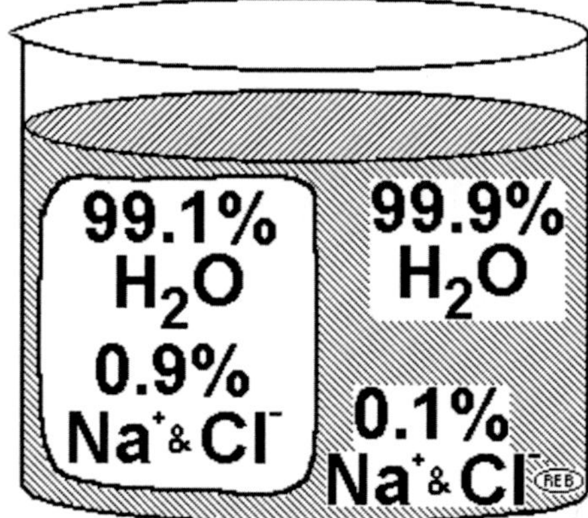

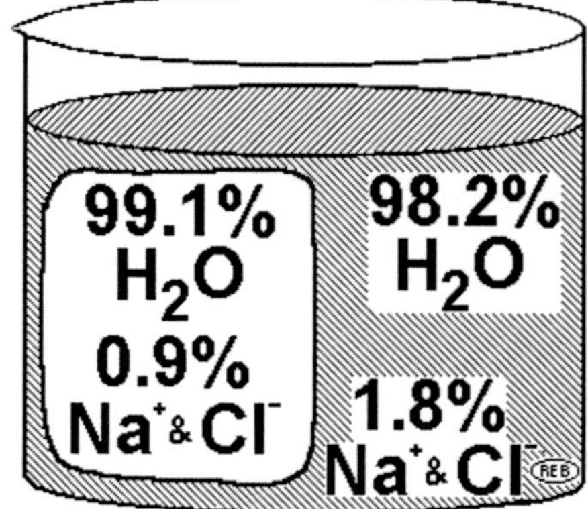

Isotonic- same concentration of solute on each the membrane. No net movement of water.

Hypotonic- lower concentration of solute than the solution on the other side of the membrane. Water moves into the cell.

Hypertonic-higher concentration of solute than the solution on the other side of the membrane. Water moves out of the cell.

The plasma membrane of the cell is a differentially permeable membrane that allows water to freely diffuse across it but limits the diffusion of solute particles. Tonicity describes the relative concentrations of solutes and solvents inside and outside of a cell relative to the cell's point of view. The environment of the cell may be described as isotonic, hypotonic, or hypertonic. In an isotonic environment the concentrations of solutes and solvents inside the cell are the same as they are outside the cell. Will there be a net movement of water into or out of a cell in an isotonic solution? Cells in a hypotonic environment are surrounded by a solution with a lower concentration of solutes than the solutes inside the cell. If the cell membrane will not allow for the movement of solutes, will water diffuse into or out of a cell in a hypotonic solution? In a hypertonic solution, the concentration of solutes is higher than the solutes in the cell. Which way would you expect water molecules to flow in a cell that is in a hypertonic solution? You and a partner are going to set up an experimental situations to answer the above questions.

Beaker	Contents of Beaker	Contents of Sealed Dialysis Tubing
1	Tap water	Tap water
2	Tap water	20 $\%_{m/v}$ Sucrose
3	20 $\%_{m/v}$ Sucrose	Tap water

1. You should be wearing your safety goggles. Pair up with another student. Obtain three 10-12 cm strip of dialysis tubing and six 10-15 cm lengths of string from the supply bench. The dialysis tubing and string may already be cut to the desired length. If dialysis tubing and string are not already cut, use the scissors provided to cut them into the appropriate lengths. Carry dialysis tubing and string back to your lab station.

2. Seal one end of each strip of dialysis tubing using the procedure that was employed in the above experiment, **B. Dialysis**. If you did not do the **B. Dialysis**, review steps **2.-4.** Above of **B. Dialysis**. Place each sealed dialysis tubing in either a 125 ml or 250 ml beaker. Trim the strings sealing the dialysis tubing so less than 1 cm remains from the knot. You and your partner should have three beakers with a closed dialysis tube in each one.

3. Carry the one of the beakers and a funnel to the supply bench. Label that beaker **1** using a grease pencil. Use your funnel and carefully transfer approximately 10 ml of tap water into the sealed dialysis tubing. Seal the open end of the dialysis tubing with string using a square knot and trim the string ends. Use a squirt bottle and gently wet the sealed ends of the dialysis tubing. Place the sealed dialysis tubing in the beaker **1** and carry it to a balance.

4. Place a mass pan on the balance and tare or zero the balance. Place the sealed dialysis tubing in the mass pan. Record the mass of the sealed dialysis tubing with tap water in the data collection table in **10.** Return the sealed dialysis tubing to its beaker and cover it with tap water. Place beaker **1** in a safe spot on your lab space.

5. Carry the second beaker and a funnel to the supply bench. Label that beaker **2** using a grease pencil. Use your funnel and carefully transfer approximately 10 ml of 20 $\%_{m/v}$ sucrose into the sealed dialysis tubing. Seal the open end of the dialysis tubing with string using a square knot and trim the string ends. Use a squirt bottle and gently wet the sealed ends of the dialysis tubing. Place the sealed tubing with in beaker **2** and carry it to a balance.

6. Place a mass pan on the balance and tare or zero the balance. Place the sealed dialysis tubing in the mass pan. Record the mass of the sealed dialysis tubing with 20 $\%_{m/v}$ sucrose in the data collection table in **10**. Return the sealed dialysis tubing to its beaker and cover it with tap water. Place beaker **2** in a safe spot on your lab space along with beaker **1**.

7. Carry the third beaker and a funnel to the supply bench. Label that beaker **3** using a grease pencil. Use your funnel and carefully transfer approximately 10 ml of tap water into the sealed dialysis tubing. Seal the open end of the dialysis tubing with string using a square knot and trim the string ends. Use a squirt bottle and gently wet the sealed ends of the dialysis tubing. Place the sealed tubing with in the beaker **3** and carry it to a balance.

8. Place a mass pan on the balance and tare or zero the balance. Place the sealed dialysis tubing in the mass pan. Record the mass of the sealed dialysis tubing with tap water in the data collection table in **10**. Return the sealed dialysis tubing to beaker **3** and cover it with 20 $\%_{m/v}$ sucrose solution. Place beaker **3** in a safe spot on your lab space along with beakers **1** & **2**.

9. At the end of 30 minutes, carry all three beakers to the balance. Remove the dialysis tubing from beaker **1** and hold it over the beaker for a count of three. Place a mass pan on the balance and tare the balance so it reads zero. Place the dialysis tubing in the mass pan and record the final mass of the dialysis tubing from beaker **1** in the data collection table in **10**. Follow the same procedure to determine the final masses of the dialysis tubing in beakers **2** & **3**.

10. Record the initial and final masses of the three sealed dialysis tubing results in the data table below. All masses should be to the 0.01 g.

Beaker	Final mass, g	Initial mass, g	Mass change, g (F– I)	Solution in beaker as compared to sealed dialysis tubing (isotonic, hypotonic, or hypertonic)
1				
2				
3				

11. Subtract the initial mass from the final mass and record the difference in the fourth column of your data table. Note a positive value indicates that the dialysis tubing gained water and a negative value indicates a loss of water. If the absolute value percent change is less than or equal to 5.00%, then it is within acceptable limits of experimental error and you can conclude that any change in mass is insignificant and report that no change has occurred. The calculation of experimental error is:

$$\% \, error \; = \; \left(\frac{change \, in \, mass}{original \, mass} \right) 100$$

The absolute value percent change in the hypotonic beaker and hypertonic beaker should be greater than 5.00%. If the experimental error is greater than 5.00% for the isotonic beaker, you may repeat the experiment if you have time and be more careful when determining the mass of your isotonic bag.

12. Was the change of mass of the dialysis tubing in the isotonic beaker significant? What was the experimental error?

____

13. Was the change of mass of the dialysis tubing in the hypotonic beaker significant? What was the experimental error?

____

14. Was the change of mass of the dialysis tubing in the hypertonic beaker significant? What was the experimental error?

____

15. Return your lab tray to its cabinet.

Notes:

Name: _______________________________

crn: _______________________________

Outcomes
Distinguish between solutions, colloids, and suspensions.
Solve problems involving concentration of solutions.

A. Prelab Assignment

The concentration of a solution is a description of how much solute is dissolved in the solvent. You are going to learn about three types of concentration- 1. Percent volume to volume, $\%_{v/v}$, 2. Percent mass to volume, $\%_{m/v}$, and 3. Molarity, M. When the solvent is not mentioned, it is understood to be water.

1. %v/v- Percent volume to volume solutions are used when a solute and solvent are both liquids. A $\%_{v/v}$ solution is the ratio of the volume of the solute to the volume of the solution. The general formula is:

$$\%_{v/v} = \left(\frac{\text{mL of solute}}{\text{mL of solution}} \right) 100$$

As an example, how would you make 250.0 ml of an 18.0 %v/v C_2H_5OH (ethanol) solution? A 100 ml solution of an 18 %v/v C_2H_5OH solution would have 18.0 ml of ethanol. Since you want to make a 250 ml solution, you will need 2.5 times 18 ml or 45.0 ml. Measure out 45.0 ml of ethanol and add enough water to make up exactly 250.0 ml of solution. An alternative method is to use ratio and proportion.

$$\frac{18 \text{ mL } C_2H_5OH}{100 \text{ mL solution}} = \frac{X}{250 \text{ mL solution}}$$

$$X = \frac{18\,(250)}{100} \text{ mL } C_2H_5OH$$

$$X = 45.0 \text{ ml } C_2H_5OH$$

With either calculation:

$$18\,\%_{v/v}\, C_2H_5OH = \left(\frac{45 \text{ mL } C_2H_5OH}{250 \text{ mL solution}} \right) 100$$

How many ml of acetone, C_3H_6O, would you need to make 25.0 ml of 4.0$\%_{v/v}$ solution of acetone in water? Show your calculations to receive credit.

1. _______________ ml C_3H_6O

2. %m/v- A percent mass to volume solution is a ratio of the mass solute to the volume of solution. The general formula is:

$$\%_{m/v} = \left(\frac{\textbf{grams of solute}}{\textbf{mL of solution}} \right) 100$$

Consider, how would you make a 1.00 liter of a 0.90 %m/v NaCl of solution? A 100 ml solution of 0.90 %m/v NaCl of solution would have 0.90 grams of NaCl. Since you are making a 1 liter or 1000 ml solution, you will need 10 times 0.90 or 9.0 grams. Mass out exactly 9.0 grams of NaCl and add quantity sufficient water to make up exactly 1.00 liter of solution. As before ratio and proportion may be used.

$$\frac{0.90 \text{ g NaCl}}{100 \text{ mL solution}} = \frac{X}{1000 \text{ mL solution}}$$

$$X = \frac{0.90\,(1000)}{100} \text{ g NaCl}$$

$$X = 9.0 \text{ g NaCl}$$

With either calculation:

$$0.9\,\%_{m/v} \textbf{ NaCl} = \left(\frac{\textbf{9.0 g NaCl}}{\textbf{1000 mL solution}} \right) 100$$

How many grams of glucose, $C_6H_{12}O_6$, would you need to make 500 ml of 5.00 %m/v solution of glucose in water? Show your calculations to receive credit.

2. _______________ g $C_6H_{12}O_6$

3. Molarity is defined as the number of moles of solute per liter of solution. So a 1 molar (M) solution has exactly one mole of the solute dissolved in exactly one liter of solution. The general formula for molarity is:

$$\textbf{Molarity} = \frac{\textbf{moles of solute}}{\textbf{liters of solution}} = \textbf{M}$$

In making a molar solution, you need to know three things- the desired molarity, the desired volume in liters, and the gram-molar mass of the solute. How would you make a 2.00 liter of a 3.00 M $Ca(NO_3)_2$ solution? The desired molarity is 3.00 M, the desired volume is 2.00 liters, and the formula mass of $Ca(NO_3)_2$ is 110.98 amu so the calculation is:

Molar solution = Desired molarity x Desired volume in liters x Gram-molar mass of solute

$$3.00 \text{ M } Ca(NO_3)_2 = 3.00 \times 2.00 \times 110.98 \text{ g } Ca(NO_3)_2 = 665.88 \text{ g } Ca(NO_3)_2$$

To make the solution, you would dissolve 665.88 g of $Ca(NO_3)_2$ with enough water to make a 2.00 liter solution.

4. Before lab, complete **B. Mixture Preparation 1a, 2a, 3c,** and **4a**.

B. Mixture Preparation

!!! SAFETY GOGGLES MUST BE WORN DURING ALL MIXTURE PREPARATIONS!!!

You will now prepare four mixtures that you will observe and then classify as either a solution, a colloid, or a suspension.

1. Preparation of Fe and water mixture

a. The Fe mixture will be a 10.0 milliliter of 1.0 %m/v mixture. Calculate the amount of Fe you will need to mix with water to make the mixture.

b. Measure out the amount of Fe from your calculations using a balance and a mass boat. Carry the mass boat with the Fe back to your lab station.

c. Set a clean test tube in a test tube rack. Add about 3-4 ml of H_2O to a 25.0 ml graduated cylinder.

d. Carefully transfer the Fe from the mass boat to the 25.0 graduated cylinder with the H_2O. Use your glass stirring rod and gently stir the mixture of Fe and the H_2O until it is completely mixed.

e. Use a disposable pipette to add enough H_2O to bring the volume of the mixture to exactly 10.0 ml. Stir the mixture to ensure a complete mixing.

f. Transfer the contents of the graduated cylinder to the clean test tube, mark the test tube as #1, and set aside it aside in a test tube rack.

2. Preparation of NaCl and water mixture

a. The NaCl mixture will be a 10.0 milliliter of 0.25 M NaCl mixture. Calculate the amount of NaCl you will need to mix with water to make the mixture.

b. Measure out the amount of NaCl from your calculations using a balance and a mass boat. Carry the mass boat with the NaCl back to your lab station.

c. Set a clean test tube in a test tube rack. Add about 3-4 ml of H_2O to a 25.0 ml graduated cylinder.

d. Carefully transfer the NaCl from the mass boat to the 25.0 graduated cylinder with the H_2O. Use your glass stirring rod and gently stir the mixture of NaCl and the H_2O until it is completely mixed.

e. Use a disposable pipette to add enough H_2O to bring the volume of the mixture to exactly 10.0 ml. Stir the mixture to ensure a complete mixing.

f. Transfer the contents of the graduated cylinder to the clean test tube, mark the test tube as #2, and set aside it aside in the test tube rack with the first test tube.

3. Preparation of glucose and water mixture

a. The glucose mixture will be the same concentration as the NaCl mixture. The glucose mixture will be made by dilution from a 20.0%$_{m/v}$ glucose stock mixture. Dilutions are based on the principle that there is a negative proportional relationship between concentration and volume. If the volume of a mixture is increased by adding more solvent, then the concentration of the mixture will decrease. This can be summed up by the expression:

$$C_1 \ x \ V_1 \ = \ C_2 \ x \ V_2$$

Where C_1 is the initial concentration of the stock mixture, V_1 is the amount of the stock mixture to be diluted, C_2 is the final concentration desired, and V_2 is the desired volume of your final mixture. The concentrations and volumes must be in the same units in the above equation.

b. A 0.25 M glucose mixture would have 0.25 mole of glucose per liter of mixture. Since the formula mass of glucose is 180.16 amu, 0.25 mole of glucose would be 45.04 grams of glucose per liter of mixture. Since 100 ml of a 0.25 M glucose mixture would have one tenth of 45.04 g or 4.50 g (0.1 x 45.04g) of glucose then a 0.25 M glucose mixture would be a 4.5 %$_{m/v}$ glucose mixture.

c. Calculate the amount of 20.0 %$_{m/v}$ glucose stock mixture you will need to mix with water to make a 10.0 ml of 0.25 M glucose mixture.

d. You will be using a disposable pipette to measure and transfer the amount of stock mixture needed. The disposable pipettes have four marks on them. Each mark is 0.25 ml so the highest mark is exactly 1.0 ml.

e. Use a disposable transfer pipette to add the amount of 20.0 %$_{m/v}$ glucose stock mixture from your calculations to a 25.0 ml graduate cylinder. Add quantity sufficient H_2O to bring the mixture's volume to 10.0 ml.

f. Transfer the contents of the graduated cylinder to the clean test tube, mark the test tube as #3, and set aside it aside in the test tube rack with the other two test tubes.

4. Preparation of starch and water mixture

    a. The starch mixture will be a 10.0 ml of 1.0 $\%_{m/v}$ starch mixture. To ensure that you have ten ml of mixture, you will be making a 20.0 ml of 1.0 $\%_{m/v}$ starch mixture. Calculate the amount of starch you will need to mix with water to make the mixture.

    b. Measure out the amount of starch from your calculations using a balance and a mass boat. Carry the mass boat with the starch back to your lab station.

    c. Get a hot plate and plug it in at your lab bench. Make sure the hot plate is off. Place a 50 ml beaker on the hot plate and add enough H_2O to cover the bottom of the beaker.

    d. Carefully transfer the starch from the mass boat to the 50 ml beaker on the hot plate. Add enough H_2O to bring the volume of the mixture to the 20 ml mark.

    e. Turn the hot plate to high and use a stirring rod to gently swirl the mixture on the hot plate. Continue swirling the mixture as the starch dissolves and the mixture warms up. Continue to swirl the mixture until you begin to see bubbles forming at the bottom of the beaker. As soon as you see any bubbles lower the heat and continue to stir for about a minute. **DO NOT ALLOW THE MIXTURE TO BOIL.**

    f. Place a clean test tube in the test tube rack with the first three test tubes. Use the contents of the first three test tubes, which are both 10.0 ml, as a guide and transfer 10.0 ml of the contents of the beaker to the third test tube, mark the new test tube as #4, and set aside it aside in the test tube rack with your first three mixtures. Take your test tube rack with your four solutions to your instructor.

C. Mixture Observation

!!! SAFETY GOGGLES MUST BE WORN DURING ALL MIXTURE OBSERVATIONS!!!

    A solution is a homogeneous mixture of a solute evenly distributed in a solvent. The particles of the solute do not settle on standing nor will any layers form in a solution. A colloid is also a homogeneous mixture but consists of two separate phases- the dispersed phase and the continuous phase. Milk is colloid of fat globules (dispersed phase) and water (continuous phase). The water in milk also has dissolved proteins. So milk is a solution of proteins dissolved in water and a colloid of fat dispersed among the water. Colloids generally appear cloudy and translucent and will scatter light. Solutions do not scatter light. The scattering of light is described as the Tyndall effect. A suspension is a mixture of a particles and a solvent but the particles of a suspension will settle out upon on standing. Blood is a suspension of blood cells in plasmas, the liquid component of blood. Plasma is a solution of water and dissolved solutes. The blood cells will settle upon standing. When a suspension is stirred the particles are evenly distributed and will scatter light. As the suspension is left to stand, the particles settle out and the liquid phase will not exhibit the Tyndall effect. Suspensions are tested for the Tyndall effect after they have been allowed to stand. If a mixture is positive for the Tyndall effect but has settled particles, it has not set for a sufficient amount of time. The differences among solutions, colloids, and suspensions are summarized below.

BIO1100
Lab 5 Solutions

	Solution	Colloid	Suspension
Separates on standing	No	No	Yes
Tyndall Effect	No	Yes	No

1. Observe your four mixtures. Use a flashlight and shine light through each solution. Which ones exhibit the Tyndall effect? Which ones have particles that have settled? Fill out the first two columns of the table below and then classify each of your mixtures as a solution, colloid, or suspension.

Mixture	Separates on standing	Tyndall Effect	Solution, Colloid, or Suspension
1. Fe			
2. NaCl			
3. Glucose			
4. Starch			

2. Dispose of your solutions, colloids, and suspensions down the sink. Flush the sink with copious amounts of water for at least one minute.

D. Polarity

!!! SAFETY GOGGLES MUST BE WORN DURING ALL POLARITY OBSERVATIONS!!!

Recall that polarity describes the degree of electron sharing in the molecules of a compound. Molecules that have equal to slightly less than equal sharing of electrons among their atoms are nonpolar. In polar molecules, the degree of electron sharing is much less than equal. Realize that polarity is a continuum so some molecules are described as more nonpolar than polar or more polar than nonpolar. Nonpolar solvents dissolve nonpolar solutes and polar solvents dissolve polar and ionic solutes. The closer the polarities of the solvent and the solute, the better the solvent dissolves the solute.

1. Arrange 14 clean test tubes in two rows of 7 in a test tube rack. Fill each test tube of one row with about two ml of H_2O. Water is a very polar solvent. In your test tubes, one finger width is about two ml. The exact amount is not as critical as the amounts being equal.

2. Fill each test tube of the other row with about two ml of cyclohexane. Cyclohexane is a very nonpolar solvent. The amount of cyclohexane in these tubes should be the same amount as water in the first set of tubes. Dispose of test solutions containing cyclohexane, iodine, or pentane in the waste bottle.

3. Use the flat end of your spatula and add a small amount of NaCl to one of the test tubes containing H_2O and stir the mixture with your glass stirring rod. Clean your spatula and stirring rod.

48

4. Use the flat end of your spatula and add a small amount of NaCl to one of the test tubes containing cyclohexane and stir the mixture with your glass stirring rod. Clean your spatula and stirring rod.

5. Compare the two tubes. Which solvent, the polar water or the nonpolar cyclohexane, did the best job of dissolving the NaCl solute? Mark your answer in the data table below. Mark an **S** for soluble and an **I** for insoluble. A solute that dissolves in a polar solvent will not dissolve in a nonpolar solvent.

6. Repeat the above procedure with the following solutes: $CuSO_4$, I_2, $KMnO_4$, pentane, oleic acid, and sucrose. If there is a case where a solute appears to dissolve in both solvents, mark the solute as soluble in the solvent that best dissolves the solute as soluble. Use a disposable transfer pipette to transfer about one ml of the liquid solutes to the test tubes. Remember, dispose of test solutions containing cyclohexane, iodine, or pentane in the waste bottle. All other test solutions can be poured down the drain.

	Solvent	
Solute	**Water**	**Cyclohexane**
Sodium chloride, NaCl		
Copper II sulfate, $CuSO_4$		
Iodine, I_2		
Potassium permanganate, $KMnO_4$		
Oleic acid, $C_{18}H_{34}O_2$		
Pentane, C_5H_{12}		
Sucrose, $C_{12}H_{22}O_{11}$		

7. Remember, polar solvents dissolve polar and ionic solutes and nonpolar solvents dissolve nonpolar solutes, "Like dissolves like." So the solutes that dissolve in water are either polar or ionic compounds and the solutes that dissolve in cyclohexane are nonpolar. Write the names or symbols of the solutes in the data table in the appropriate blanks below.

Polar or ionic solutes ____

____

____

Nonpolar solutes ____

____

____

8. Return your lab tray to its cabinet.

Notes:

BIO1100
Lab 6 Electrolytes

Name: _______________________________

crn: _______________________________

<table>
<tr><td align="center">Outcomes
Apply knowledge of buffers to physiological systems
Identify important physiological electrolytes
Describe the properties of acids and bases; use the pH scale</td></tr>
</table>

A. Prelab Assignment

Use printed or internet sources to answer the questions below.

1. Acid: _______________________________

____

____

____

1. Base: _______________________________

____

____

____

2. Buffer _______________________________

____

____

____

3. Electrolyte: _______________________________

____

____

____

4. Equivalent: _______________________________

____

____

____

B. Conductivity

Electrolytes are substances with partial to complete dissociation or ionization in water. Salts, acids, & bases are three categories of electrolytes. Electrolytes are described as strong to weak depending on the degree of ionization they exhibit. A strong electrolyte has complete to high dissociation in aqueous solutions. Complete to high dissociation results in high ion concentration which results in strong electrical conductivity. $NaCl$, $CuSO_4$, HCl, $NaOH$, and $KMnO_4$ are all examples of strong electrolytes. HF, H_2O, NH_4OH, H_3BO_3, and $HC_2H_3O_2$ are all examples of weak electrolytes. Weak electrolytes have limited dissociation or ionization in aqueous solutions, producing few ions, and, therefore, are poor conductors. Nonelectrolytes do not dissociate to form ions and are nonconductors. Sugars, alcohols, oils, and most organic compounds are nonelectrolytes. The conductivity differences among strong electrolytes, weak electrolytes, and nonelectrolytes are summarized below.

The degree of ionization can be described as a percent. Both HCl and HNO_3 are strong acids and both have 92% ionization. This means in the case of HCl, at any given moment in a solution of HCl, 92% of the hydrochloric acid exists as H^+ and Cl^- ions while 8% is in molecular form as HCl. This is an equilibrium so as some of the H^+ and Cl^- ions are uniting to form HCl, some of HCl molecules are dissociating to form H^+ and Cl^-. Acetic acid, $HC_2H_3O_2$, is a weak acid and has a percent ionization of 1.3%, so at any given moment there are 1.3% H^+ and $C_2H_3O_2^-$ ions and 99.7% $HC_2H_3O_2$ molecules. Sugars and alcohols are nonelectrolytes. Nonelectrolytes do not ionize, so their percent ionization is zero.

Electrical Properties of Electrolytes		
Substance	**Conductivity**	**Composition**
Strong Electrolytes	Strong	Mostly ions
Weak Electrolytes	Weak	Mix of ions & molecules
Nonelectrolytes	None	Molecules

!!! SAFETY GOGGLES MUST BE WORN DURING ALL CONDUCTIVITY OBSERVATIONS!!!

1. Obtain a porcelain spot plate and a conductivity testing apparatus from the supply bench and take them back to your lab space.

2. Place a sample of each substance in a separate well of the spot plate in the order that the substances appear in the data collection table below. Fill a 250 mL beaker with RO water to the 200 mL mark.

3. Place the electrodes of the conductivity testing apparatus into the first sample. Note and record the brightness of the light by placing an X in one of the three columns of the data collection table. Rinse the electrodes by dipping them into the RO water. Wipe the electrodes off with a paper towel.

4. Repeat this procedure with the rest of the samples. Be sure to rinse and wide the electrodes after testing each sample.

Substance	Strong Electrolyte (bright light)	Weak Electrolyte (glowing light)	Non-Electrolyte (no light)
RO water			
Tap water			
Sucrose, $C_{12}H_{22}O_{11}$			
Ethanol, C_2H_5OH			
Sulfuric acid, H_2SO_4			
Sodium hydroxide, NaOH			
Ammonium hydroxide, NH_4OH			
Boric acid, H_3BO_3			
Sodium chloride, NaCl			
Copper II sulfate, $CuSO_4$			
Cyclohexane, C_6H_{12}			
Acetic acid, $HC_2H_3O_2$			

5. Which of the above substances are strong electrolytes?

____

____

____

6. Which of the above substances are weak electrolytes?

____

____

____

7. Which of the above substances are non-electrolytes?

____

____

____

C. Indicators

Acids and bases are electrolytes that are not salts. Acids release H^+ in water and form H_3O^+ (hydronium ion). The third hydrogen in H_3O^+ readily leaves the hydronium ion, therefore acids are proton donors. Bases release OH^- (hydroxide ions) in water. The hydroxide ions will readily combine with protons and are proton acceptors. The strength of acids and bases using the pH scale. The pH scale is based on the ionization of water. The electrons that are being shared between the hydrogen atoms and the oxygen atom in water are unequally shared. The sharing is so unequal that H_2O molecules will break apart to form H^+ and OH^-. The H^+ will quickly take up residence at the negative area around water to form H_3O^+. Since the hydronium ion is only transitory, the ionization of water can be shown with hydrogen ion, H^+ as a product as shown, below.

$$H_2O \rightleftharpoons H^+ + OH^-$$

Many chemical reactions go to completion, that is to say all the reactant or reactants are changed to product or products. These reactions are called forward or right reactions. In many reactions, only some of the reactant becomes product and while reactant is becoming product some of the product is going back to being reactant. These reactions are called back or left reactions. There is an equilibrium between the reactant and product.

A measure of the balance of reactants and products in an equilibrium reaction is the equilibrium constant, K_{eq}. The equilibrium equation is a ratio of the molar concentration of the products to the molar concentration of the reactants. In the reaction below, the brackets stand for molar concentration.

$$K_{eq} = \frac{[Products]}{[Reactants]}$$

If a reaction is mostly forward and most of the reactants are changed to product, then the K_{eq} will be large. If only a small amount of product is produced, then the K_{eq} will be small. The equilibrium constant for the ionization of water is called K_w and is

$$K_w = \frac{[H^+][OH^-]}{[H_2O]}$$

For one mole of water, one ten millionth of a mole of H^+ and OH^- are produced so the equilibrium constant for water is

$$K_w = \frac{[10^{-7}][10^{-7}]}{[1]} = 10^{-14}$$

Note that if the concentration of H^+ increased to 10^{-3}, the OH^- concentration would then decrease to 10^{-11} to maintain the K_w of 10^{-14}. This relationship is the basis for the definition of pH and the pH scale. The abbreviation pH is short for "power of Hydrogen" and is defined as

$$pH = -\log [H^+]$$

So if the hydrogen ion concentration of a solution is 10^{-3} moles, the pH of the solution is

$$pH = -\log 10^{-3} = -(-3) = 3$$

Notice as the hydrogen ion concentration of a solution increases, the pH value decreases. A solution with a pH of 11 would have a very low concentration of hydrogen ions. The pH scale ranges from 0 to 14. Solutions with a pH greater than 0 but less than 7 are acids or acidic, solutions with a pH of exactly 7 are neutral, and solutions with a pH greater than 7 but less than 14 are bases or basic.

pH Scale
0   < pH <   7   < pH <   14
Acids   ↑   Bases
Neutral

The concentration of OH⁻ ions of a solution can also be measured and the pOH calculated but it is generally not done. The pH and the pOH of a solution must equal 14, so if the pH of a solution is 11, then the pOH would be 3.

$$pH + pOH = 14$$

!!! SAFETY GOGGLES MUST BE WORN DURING ALL INDICATOR OBSERVATIONS!!!

There are many methods to determine the pH of a solution. Some are more precise than others. Some indicators can only tell you if a solution is acidic or basic, some methods can measure pH to the ones place, and some instruments can determine pH to the tenths place or hundredths place or beyond.

1. Litmus paper is an extract obtained from several species of lichens. Acids turn blue litmus red and bases turn red litmus blue. In acids, red litmus stays red and blue litmus stays blue in bases.

   a. Arrange 10 clean test tubes in a test tube rack in two rows of five. Mark one row of test tubes with an A and mark the other row with a B.

   b. Obtain Unknown A and a transfer pipette from the lab supply bench. Use caution, Unknown A is either an acid or a base. Use the transfer pipette and carefully transfer 2 mL of Unknown A into each one of the A row test tubes. Two mL is about two full transfer pipettes. Return Unknown A to the supply bench. Repeat this procedure with Unknown B placing 2 mL of Unknown B into each of the B row test tubes.

   c. Obtain two strips of red litmus paper and two strips blue litmus paper from the supply bench and take them back to your lab bench. Place a strip of red litmus and a strip of blue litmus in the first A row test tube and place a strip of red litmus and a strip of blue litmus in the first B row test tube. Gently swirl each tube so that the litmus strips are completely saturated by the unknown solutions.

   d. In each tube, one strip will change color but one strip will not. Both strips in a test tube will be the same color. Record the color of the strips in the A and B test tubes in the data table below as either "red" or "blue."

   e. Empty to liquid contents of each test tube into the lab sink but do not empty of the Litmus paper into the sink. Dispose of the litmus paper in a waste container.

2. Phenolphthalein is often abbreviated "phph" and remains clear and colorless in acid solutions but turns reddish purple in base solutions.

   a. Obtain a container of phenolphthalein and a transfer pipette from the lab supply bench.

   b. Use the transfer pipette to carefully add 2-3 drops of phenolphthalein to an A row test tube and then add 2-3 drops of phenolphthalein to a B row test tube. Gently swirl each test tube and note the appearance of the contents of each.

   c. Record the appearance of the A and B test tube contents in the data table below as either "clear & colorless" or "red-purple." Dispose of the contents of each test tube in the lab sink.

3. Methyl orange is an acid-base indicator that turns from bright orange to reddish pink in acid solutions and yellowish orange or straw yellow in base solutions.

   a. Obtain a container of methyl orange and a transfer pipette from the lab supply bench.

   b. Use the transfer pipette to carefully add 2-3 drops of methyl orange to an A row test tube and then add 2-3 drops of methyl orange to a B row test tube. Gently swirl each test tube and note the appearance of the contents of each.

   c. Record the appearance of the A and B test tube contents in the data table below as either "reddish pink" or "straw yellow." Dispose of the contents of each test tube in the lab sink.

4. There are many brands of pH paper. All of them have a multi-range indicator that turns a distinct color at a specific pH value. By comparing the resulting color to a chart, pH can be determined to the ones place.

   a. Obtain two strips of pH paper from the supply bench and take them back to your lab bench. Place a one strip of pH paper in the an A row test tube and place the other strip of pH paper in a B row test tube. Gently swirl each tube so that the indicator portion of the pH strips is completely saturated by the unknown solutions.

   b. Observe the color of the indicator portion of each strip and match the resulting color to a number on the pH paper chart. Record the number on the chart in the data table below.

   c. Empty to liquid contents of each test tube into the lab sink but do not empty of the pH paper into the sink. Dispose of the pH paper in a waste container.

5. A pH meter uses the conductivity of a solution to measure the pH of the solution. While a pH meter senses the difference in electric potential between positive and negative electrodes, a processor changes the voltage measurement into a value on the pH scale that can then be read on a digital display.

   a. Obtain a pH meter from the lab supply bench. Read the directions on the pH meter and follow them carefully.

   b. Place the electrodes of the pH meter in the last A test tube. Record the reading in the data table. Hold the electrodes of the pH meter over the sink and rinse the electrodes off using RO water. After the electrodes are well rinsed, place them in the last B test tube. Record the reading in the data table and rinse off the electrodes. Empty to contents of the test tubes into the lab sink.

Indicator	Unknown A	Unknown B
Litmus paper		
Phenolphthalein		
Methyl orange		
pH paper		
pH meter		

6. Which of the above indicators is the most precise?

____

7. Based on your observations, is unknown A an acid or a base?

____

8. Based on your observations, is unknown B an acid or a base?

____

9. Is a solution more acidic at 3 pH or 6 pH?

____

10. Is it possible for a solution that is not 7 pH to be neutral?

____

11. Is a base strong or weak at a high pH?

____

12. The acid-base balance of a solution is 11 pH, what is its pOH?

____

13. Is the above solution acidic or basic?

____

14. Methanoic acid (formic acid) has a percent ionization of 4.2%, what percent of methanoic acid is in molecular form in an aqueous solution?

____

15. Return your lab tray to its cabinet.

Notes:

58

Name: ___________________________

crn: ___________________________

Outcomes
List the major classes of organic molecules. Using the International Union of Pure and Applied Chemistry (IUPAC) nomenclature, properly name simple organic compounds.

A. Prelab Assignment

Organic chemistry involves covalently bonded reduced carbon compounds. The most reduced carbon compounds are the hydrocarbons. Hydrocarbons contain only hydrogen and carbon. Hydrocarbons are named by the number of carbon atoms and by the type of covalent bonds present. Covent bonds involve the sharing of electron pairs. In a single covalent bond, a pair of electrons (two electrons) is shared between two atoms. Double covalent bonds have two pairs of electrons (four electrons) bonding two atoms together. How many electron pairs do you think would be involved in a triple covalent bond?

Alkanes are hydrocarbons that that have only single covalent bonds. The simplest hydrocarbon is methane with only one carbon and four single covalent bonds holding the four separate hydrogen atoms to the central carbon. Ethane is a hydrocarbon with two carbons covalently bonded to each other and with each carbon holding three covalently bonded hydrogens. Propane is a chain of three carbons with the two end carbons each having three covalently bonded hydrogens and the middle carbon having two covalently bonded hydrogens. Below are full structural formulae of methane, ethane, propane, and butane. Each line represents a pair of shared electrons. Circle the structural formula of propane.

1.

$$\begin{array}{cccc} H & H & H & H \\ | & | & | & | \\ H-C-C-C-C-H \\ | & | & | & | \\ H & H & H & H \end{array}$$

$$\begin{array}{cc} H & H \\ | & | \\ H-C-C-H \\ | & | \\ H & H \end{array}$$

$$\begin{array}{ccc} H & H & H \\ | & | & | \\ H-C-C-C-H \\ | & | & | \\ H & H & H \end{array}$$

$$\begin{array}{c} H \\ | \\ H-C-H \\ | \\ H \end{array}$$

Alkenes are hydrocarbons that contain at least one double covalent carbon to carbon bond. In a double covalent bond, two pairs of electrons, four total, are being shared between two carbon atoms. The simplest alkene is ethene. Ethene has two carbon atoms with a double and each carbon has two hydrogen atoms in single covalent bonds. Propene is a three carbon chain with two of the carbon atoms sharing two pairs of electrons and a third carbon in a single covalent bond with one of the two in the double bond. The carbon with the single bond has three covalently bound hydrogen. The middle carbon has a single hydrogen and the other end carbon has two hydrogens. Butene is a four carbon chain with one double covalent bond between two of the carbons. Butene can be either 1-butene or 2-butene. In 1-butene, the double covalent bond is between an end carbon, the first carbon, and the next carbon, the second carbon, in the chain. The end carbon has two hydrogens, the next carbon has one hydrogen, the third carbon has two hydrogens, and the last carbon has three hydrogens. In 2-butene, the double bond is between the two middle carbons. The two end carbons have three hydrogens and each carbon in the double bond has one hydrogen. The hydrogen atoms attached to the double bond carbons in 2-butene can either be on the same side of the double bond or on opposite sides of the double bond. The word *cis* describes the former, while *trans* describes the latter. Below are the structural formulae for propene, 1-butene, *cis*-2-butene, and *trans*-2-butene. Circle the structural formula for *cis*-2-butene.

2.

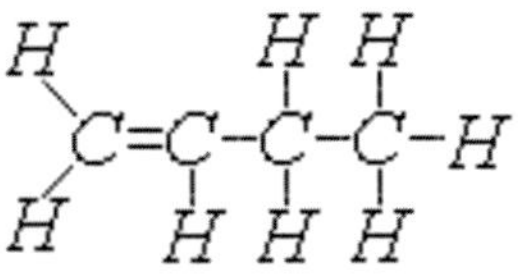

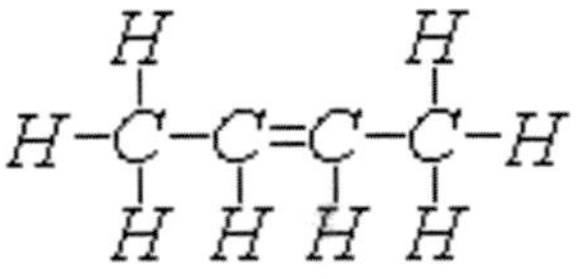

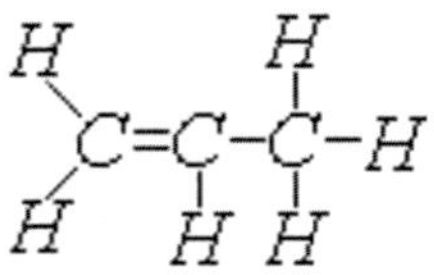

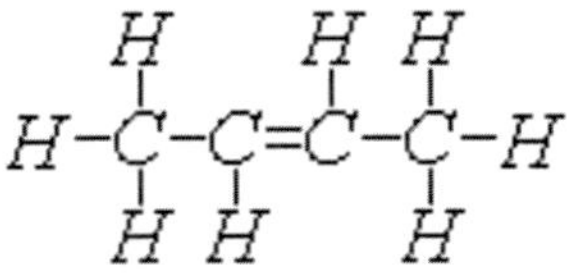

Alkynes are hydrocarbons that contain at least one triple covalent carbon to carbon bond. In a triple covalent bond, three pairs of electrons, six total, are being shared between two carbon atoms. The simplest alkene is ethyne (acetylene). Study the full structural formulae below and then circle the one representing ethyne.

3.

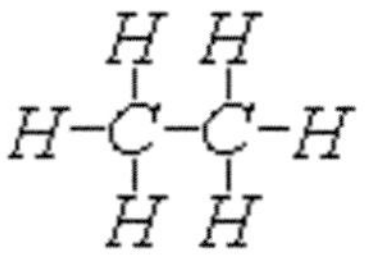

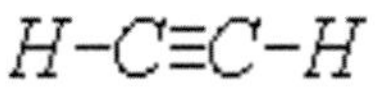

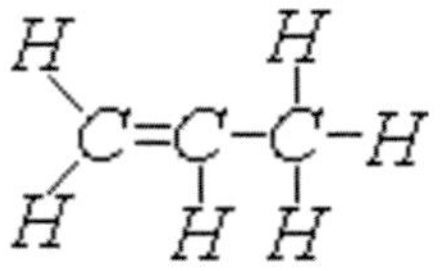

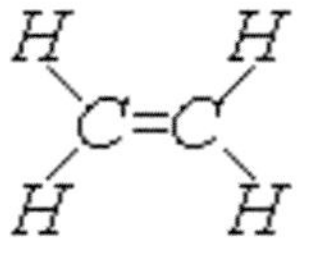

Notice that in all the structural formulae above that each carbon always forms four covalent bonds. When looking at any drawings of structural formulae, always remember that the drawings are two dimensional representations of three dimensional molecules.

B. Organic Models

!!! SAFETY GOGGLES ARE NOT REQUIRED FOR ORGANIC MODELS!!!

You are going to construct some three dimension models of representative molecules of organic compounds. Obtain a molecular model kit from the lab supply bench. Take the kit back to your lab bench and open it up. The black spheres represent carbon atoms and will simply be referred to as carbon atoms or carbons. Hydrogen atoms are generally represented by yellow or white spheres while oxygen is usually a red sphere and nitrogen a light blue sphere. Notice that there are three kinds of connectors representing covalent bonds. The short connectors are used to connect hydrogen atoms to carbon and other larger atoms such as nitrogen and oxygen. While the long connectors are used to connect carbon to carbon and carbon to nitrogen and oxygen. The spring connectors are used to form double and triple covalent bonds.

1. Ethanol (Ethyl alcohol), C_2H_5OH

a. Use a short connector and attach an oxygen to hydrogen. Next, using a short connector attach a carbon to the oxygen with the attached hydrogen.

b. Use a long connector to attach a second carbon to the first carbon that has the oxygen and hydrogen. An oxygen and hydrogen group in a covalent bond attached to carbon is a hydroxyl. Remember that an oxygen and hydrogen group in an ionic bond is a hydroxide. Hydroxides ionize but hydroxyls do not.

c. Use short connectors to attach two hydrogens to the carbon with the hydroxyl and three hydrogens to the other carbon.

d. Use the structural formulae from the prelab assignment as a guides and draw the two dimensional structural formula for ethanol in the space below. Write ethanol's molecular formula next to its structural formula. When you are finished, take your ball and stick ethanol model and your drawing to your instructor.

2. Ethyne (Acetylene), C_2H_2

a. Use three spring connectors and attach two carbons together. Next, using short connectors attach a hydrogen to each carbon.

b. Draw the two dimensional structural formula for ethyne in the space below. Write ethyne's molecular formula next to its structural formula. When you are finished, take your ethyne model and your drawing to your instructor.

3. Propanone (acetone), C_3H_6O

a. Place a long connector in a carbon. Next place short connectors in the three remaining holes of the carbon. Place a hydrogen at the end of each short connector. This configuration represents a methyl group. Think of the long connector as an electron looking for another electron to form a covalent bond. Make a second methyl group and set both the methyl groups aside.

b. Use two spring connectors and attach an oxygen to a carbon. An oxygen double bonded to a carbon is a carbonyl.

c. Attach the two methyl groups to the carbonyl carbon so that one methyl group is on each side of the carbonyl carbon.

d. Draw the two dimensional structural formula for propanone and write its molecular formula next to the structural formula. When you are finished, take your propanone model and your drawing to your instructor.

4. *trans*-2-butene, C_4H_8

a. Review the prelab assignment concerning 2-butene. Construct two methyl groups and set them aside.

b. Place a short connector into a hydrogen. Place a short connector into another hydrogen. Set the two hydrogens with short connectors aside.

c. Connect two carbons with two spring connectors. This represents a double bond. Notice that if you hold the two springs representing the double bond horizontal and flat, then the two remaining holes in each carbon are vertical.

d. Attach the one of the hydrogens to one of the double bond carbons. Attach the other hydrogen to the other double bond carbon so that the hydrogens are on opposite sides of the double bond (the springs).

e. Attach the one of the methyl groups to one of the double bond carbons. Attach the other methyl group to the other double bond carbon so that both methyl groups are on opposite sides of the double bond.

f. Draw the two dimensional structural formula for *trans*-2-butene and write its molecular formula next to the structural formula. Write the molecular formula for *trans*-2-butene next to its structural formula. When you are finished, take your *trans*-2-butene model and your drawing to your instructor.

g. When you return to your lab bench, set your *trans*-2-butene model aside. You are now going to make an isomer of *trans*-2-butene. **Isomers are compounds with the same molecular formula but with different structural formulae.**

5. *cis*-2-butene, C_4H_8

    a. Construct two hydrogens with short connectors and two methyl groups. Set them aside on your lab bench.

    b. Connect two carbons with two spring connectors. Orient the carbons so the two springs representing the double bond are horizontal and flat.

    c. This time, attach the hydrogens so that they are on different carbons and on the same side of the double bond and attach the two methyl groups so that they also are on the same side of the bond.

    d. Draw the two dimensional structural formula for *cis*-2-butene and write its molecular formula next to the structural formula. When you are finished, take your *cis*-2-butene model and your drawing to your instructor.

    e. When you return to your lab bench, compare *trans*-2-butene to *cis*-2-butene.

6. Butane, C_4H_{10}, and 2-methyl propane, C_4H_{10}

    a. Study the structural formula of butane.

$$H-\underset{\underset{H}{|}}{\overset{\overset{H}{|}}{C}}-\underset{\underset{H}{|}}{\overset{\overset{H}{|}}{C}}-\underset{\underset{H}{|}}{\overset{\overset{H}{|}}{C}}-\underset{\underset{H}{|}}{\overset{\overset{H}{|}}{C}}-H$$

    What is the molecular formula of butane? _______________________________________

    b. Construct a ball and stick model of butane and set it aside.

    c. Study the structural formula of 2-methyl propane. It is common practice in organic chemistry to draw a structural formula without showing the hydrogens. The hydrogens are understood to be present as in the drawing below.

$$\begin{array}{c}-\overset{|}{\underset{|}{C}}- \\ -\overset{|}{\underset{|}{C}}-\overset{|}{\underset{|}{C}}-\overset{|}{\underset{|}{C}}- \\ \end{array}$$

    What is the molecular formula of 2-methyl propane? _______________________________

    d. Construct a ball and stick model of 2-methyl propane and compare it to your model of butane. What name describes molecules that have the same molecular formula but different structural formulae?

____

    Remember to place the parts of your molecular back into the box the way you found them and to return the kit to the lab bench where you found it.

C. Drawing Isomers

!!! SAFETY GOGGLES ARE NOT REQUIRED FOR DRAWING ISOMERS!!!

Remember from above that **isomers are compounds that have the same molecular formula but different structural formulae.** Consider the molecular formula: C_2H_6O. This molecular formula describes two different compounds, ethanol and methoxymethane (dimethyl ether).

Ethanol	Methoxymethane (dimethyl ether)
$H-\overset{\displaystyle H}{\underset{\displaystyle H}{C}}-\overset{\displaystyle H}{\underset{\displaystyle H}{C}}-OH$	$H-\overset{\displaystyle H}{\underset{\displaystyle H}{C}}-O-\overset{\displaystyle H}{\underset{\displaystyle H}{C}}-H$

Since both compounds have the same molecular formula but different structural formulae, ethanol and methoxymethane are isomers. In ethanol, oxygen forms a single covalent bond to carbon and another single covalent bond to hydrogen. In methoxymethane, oxygen form a single covalent bond to a carbon and another single covalent bond to another carbon. Draw all the possible isomers under the molecular formulae below. Notice that oxygen can form two covalent bonds. You may want to use your molecular model kit to discover the isomers.

1. C_3H_8O (3 isomers)

2. C$_5$H$_{12}$ (3 isomers)

3. C$_4$H$_8$ (5 isomers)

Remember to place the parts of your molecular model kit back into the box the way you found them and to return the kit to the lab bench where you found it. Take your model kit to your instructor before putting it back on the lab bench.

Notes:

Name: ___________________________

crn: ___________________________

Outcomes
Define a functional group. Identify major organic functional groups and explain the properties of each. Discuss how the properties of a functional group alter the chemical properties of an organic molecule.

A. Prelab Assignment

Functional groups are a combination of atoms that chemically behave in a predictable manner. That is to say a functional group is the part of an organic molecule that largely determine the chemical properties of that molecule. Five functional groups were introduced in Lab 7 Organic Structure- alkanes, alkenes, alkynes, alcohols, and ketones. Shown below are the major functional groups of organic compounds. Use printed or internet sources to find the name of each functional group and then write the name of the functional group next to it. In the structures below, R represents a carbon chain or ring to which the functional group is attached. The first one has already been done for you.

1.
$$R-\overset{\overset{H}{|}}{C}-\overset{\overset{H}{|}}{\underset{\underset{H}{|}}{C}}-R$$
        Alkane _______________________________

2.
$$R-C=\overset{\overset{H}{|}}{\underset{\underset{H}{|}}{C}}-R$$
____

3. $R-C\equiv C-R$

____

4. $R-OH$

____

5. $R-SH$

____

6. $R-NH_2$

or

$$R-\overset{\overset{\displaystyle H}{|}}{N}-R$$

or

$$R-\overset{\overset{\displaystyle R}{|}}{N}-R$$

____

7. $R-O-R$

____

8. $R-\overset{\overset{\displaystyle O}{\|}}{C}-H$

____

9. $R-\overset{\overset{\displaystyle O}{\|}}{C}-R$

____

10. $R-\overset{\overset{\displaystyle O}{\|}}{C}-OH$

____

11. $R-\overset{\overset{\displaystyle O}{\|}}{C}-O-R$

____

12. $R-\overset{\overset{\displaystyle O}{\|}}{C}-\overset{\overset{\displaystyle H}{|}}{N}-R$

____

13.

____

BIO1100
Lab 8 Functional Groups

B. Organic Unknowns

!!! SAFETY GOGGLES ARE REQUIRED FOR ALL TEST REACTIONS!!!

Functional groups behave in predictable ways to determine the chemical properties of a compound. Knowledge of the chemical behavior of functional groups helps determine the identity of organic compounds. You will be testing three compounds – ethanol, hexane, and toluene. The compounds will be labeled X, Y, and Z. By testing the reactions of the unknowns, you will determine their names.

1. Oxidation with potassium permanganate, $KMnO_4$

a. Obtain a test tube rack from the supply bench. Label four clean test tubes (16 x 120 mm) X, Y, Z, and C. Take your test tube rack with test tubes to the fume hood. Use a transfer pipette and carefully transfer 5 mL of each unknown to the appropriate test tube. Take the test tube rack with test tubes back to your lab bench and add 5 mL of RO water to the test tube marked C. Test tube C is your control.

b. Take your test tube rack with the four test tubes to the supply bench. Using a transfer pipette carefully add 1 mL, a dropper full, of $KMnO_4$ solution to each test tube. Take the test tube rack and test tubes back to your lab bench. Place a rubber stopper in each test tube and shake each test tube vigorously. You will interpret and record the results after you perform the acid catalysis test.

c. As $KMnO_4$ oxidizes a substance, the test solution will change color and become less purple and more red to brown. A strong positive will brownish red. Compare each test tube with your control. Record the results of your $KMnO_4$ oxidation test in the data table below.

d. Dispose of all test solutions in the waste disposal bottle in the fume hood. Clean and rinse out your test tubes at your lab bench.

2. Acid catalysis with sulfuric acid, H_2SO_4

a. Obtain a test tube rack from the supply bench. Your test tubes (16 x 120 mm) must be clean Wash and rinse your test tubes and pour out as much water. Label your test tubes X, Y, Z, and C and place them in a test tube rack.

b. Take your test tube rack with test tubes to the fume hood. Use a transfer pipette and carefully transfer 1 mL of each unknown to the appropriate test tube. Add 1 mL of H_2SO_4 to the control test tube.

c. Acid catalysis produces heat. In order to record the heat of the reaction, place a thermometer in the test tube for each unknown. Note the temperature of the unknown before adding the H_2SO_4. Note the temperature of the unknown after you add 1 mL H_2SO_4 to the test tube. Record any increase in temperature as a positive reaction in the data table below.

d. Dispose of all test solutions in the waste disposal bottle in the fume hood. Clean and rinse out your test tubes at your lab bench. Remember to read and record the results of the $KMnO_4$ oxidation.

69

3. Solubility with water, H_2O

a. Label four clean test tubes (16 x 120 mm) X, Y, Z, and C. Take your test tube rack with test tubes to the fume hood. Use a transfer pipette and carefully transfer 5 mL of each unknown to the appropriate test tube. Take the test tube rack with test tubes back to your lab bench and add 5 mL of RO water to the test tube marked C. Test tube C is your control.

b. Add 1 mL of RO water to each test tube. Gently swirl each test tube to mix the contents.

c. If the contests of a test tube remain mixed, the unknown dissolved in water, and the result is positive. If the contents of a test tube separate into two distinct layers, the unknown did not dissolve in water, and the result is negative.

d. Dispose of all test solutions in the waste disposal bottle in the fume hood. Clean and rinse out your test tubes at your lab bench.

4. Data Collection

a. Record your results as + (plus sign), meaning a reaction occurred (positive result) or – (minus sign), meaning no reaction occurred (negative result). For example, if the $KMnO_4$ control did not change color, that should be recorded in your results table as a negative result with a minus sign. Unless your water control separated into two distinct layers, it should be recorded as a positive result with a plus sign.

Unknown	Oxidation with $KMnO_4$	Acid Catalysis with H_2SO_4	Solubility with H_2O
X			
Y			
Z			
Control			

5. Analysis

a. The expected result of each of your unknown compounds is shown in the table below

Compound	Oxidation with $KMnO_4$	Acid Catalysis with H_2SO_4	Solubility with H_2O
Ethanol	+	+	+
Hexane	-	-	-
Toluene	+	-	-
Control	-	-	+

b. Compare your experimental results with the expected results in the above table and then name your unknowns.

Unknown X is ____

Unknown Y is ____

Unknown Z is ____

C. Functional Group Solubility

!!! SAFETY GOGGLES ARE REQUIRED FOR ALL SOLUBILITY TESTS!!!

The solubility of organic compounds is determined by the relative amount of polar area to non-polar area of the individual molecules of each compound. Organic compounds with no polar area or a small polar area relative to their non-polar area are insoluble in water, which is a very polar solvent. Hydrocarbons are very non-polar and do not dissolve in water. As the amount of polar area relative to non-polar area increases in an organic molecule, its solubility in water increases. If the non-polar area is large relative to polar area, the compound's solubility in non-polar solvents increases. Toluene is a very non-polar solvent. Some compounds have a balanced polar to non-polar area. This allows them to dissolve in both polar and non-polar solvents. Such compounds are used as emulsifying agents.

Hydrocarbon chains are non-polar. Functional groups containing oxygen, nitrogen, sulfur, or phosphorus are polar. Polar functional groups are amides, amines, carbonyls, carboxyls, esters, hydroxyls, imines, nitriles, phosphates, and thiols.

1. Solubility of Ethanol and Octanol

a. Study the structural formulae of ethanol and octanol. Note that both have one hydroxyl group (-OH) so they each have the same amount of polar area, but that with an eight carbon hydrocarbon chain, octanol has four times the non-polar area as ethanol which only has a two carbon hydrocarbon chain.

Ethanol	Octanol
$H-\overset{\displaystyle H}{\underset{\displaystyle H}{C}}-\overset{\displaystyle H}{\underset{\displaystyle H}{C}}-OH$	$H-\overset{\displaystyle H}{\underset{\displaystyle H}{C}}-\overset{\displaystyle H}{\underset{\displaystyle H}{C}}-\overset{\displaystyle H}{\underset{\displaystyle H}{C}}-\overset{\displaystyle H}{\underset{\displaystyle H}{C}}-\overset{\displaystyle H}{\underset{\displaystyle H}{C}}-\overset{\displaystyle H}{\underset{\displaystyle H}{C}}-\overset{\displaystyle H}{\underset{\displaystyle H}{C}}-\overset{\displaystyle H}{\underset{\displaystyle H}{C}}-OH$

b. Obtain a test tube rack from the supply bench. Place four clean test tubes (16 x 120 mm) in the test tube rack so that one test tube is in the each corner of your test tube rack. Place 2 mL of RO water into the two test tubes on the left side of the test tube rack. Take your test tube rack with test tubes to the fume hood. Use a transfer pipette and carefully transfer 2 mL of toluene into the two empty test tubes on the right side of the test tube rack

c. Use a new transfer pipette and transfer 2 mL of ethanol into the upper left test tube containing water and 2 mL of ethanol into the upper right test tube containing toluene. Use a new transfer pipette and transfer 2 mL of octanol into the lower left test tube containing water and 2 mL of octanol into the lower right test tube containing toluene.

d. Vigorously shake, but do not splash or spill, the contents of all four test tubes. Allow the contents of each test tube to set for a few minutes. If the two substances are soluble (S), then two compounds will evenly distribute with each other and form a solution with no layers. If the two compounds are insoluble (I), then the two compounds will not stay mixed and will separate into two distinct layers.

e. **Data collection-** Each empty space in the table below represents one of the test tubes. The upper left space represents the test tube containing ethanol and water, the lower left square represents the octanol and water test tube, and so on. Mark each space with an S or an I as appropriate.

	Water (Polar)	**Toluene (Non-polar)**
Ethanol		
Octanol		

f. **Analysis-** Explain your results in terms of polar and non-polar area.

____

____

____

____

____

____

____

____

____

____

____

____

____

2. Solubility of Cyclohexanol and Glucose (Dextrose)

a. Study the structural formulae of cyclohexanol and glucose. Note that both have the same amount of non-polar area, the six carbon hydrocarbons but glucose has six polar groups (5 hydroxyls & 1 ether) while cyclohexanol has only one hydroxyl group.

Cyclohexanol	Glucose
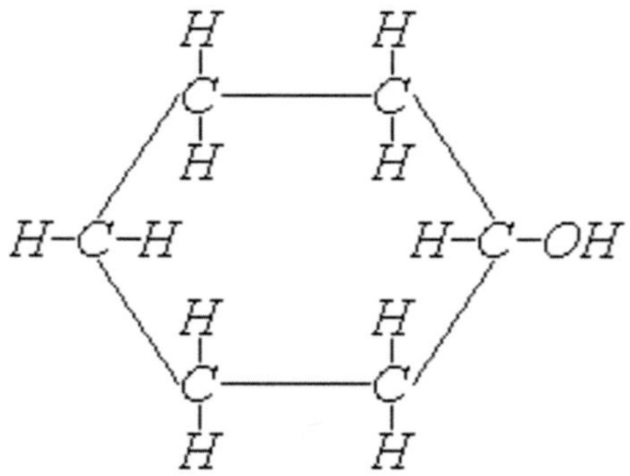	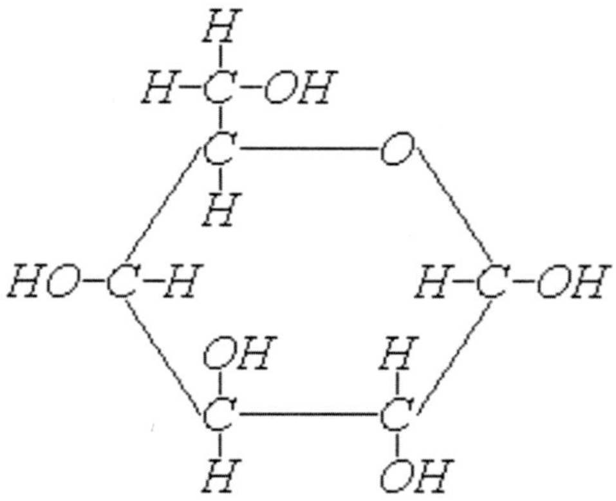

b. Obtain a test tube rack from the supply bench. Place four clean test tubes in the test tube rack so that one test tube is in the each corner of your test tube rack. Place 2 mL of RO water into the two test tubes on the left side of the test tube rack. Take your test tube rack with test tubes to the fume hood. Use a transfer pipette and carefully transfer 2 mL of toluene into the two empty test tubes on the right side of the test tube rack

c. Use a new transfer pipette and transfer 2 mL of cyclohexanol into the upper left test tube containing water and 2 mL of cyclohexanol into the upper right test tube containing toluene. Use a spatula and transfer a small amount, just cover the end of the spatula, of glucose into the lower left test tube containing water and the same amount of glucose into the lower right test tube containing toluene.

d. Vigorously shake, but do not splash or spill, the contents of all four test tubes. Allow the contents of each test tube to set for a few minutes. If the two substances are soluble (S), then two compounds will evenly distribute with each other and form a solution with no layers. If the two compounds are insoluble (I), then the two compounds will not stay mixed and will separate into two distinct layers.

e. **Data collection-** Each empty space in the table below represents one of the test tubes. The upper left space represents the test tube containing cyclohexanol and water, the lower left square represents the glucose and water test tube, and so on. Mark each space with an S or an I as appropriate.

	Water (Polar)	Toluene (Non-polar)
Cyclohexanol		
Glucose		

f. **Analysis-** Explain your results in terms of polar and non-polar area.

____

____

____

____

____

____

____

____

____

____

____

____

____

3. Return your lab tray to its cabinet.

4. Identifying Polar Functional Groups- study the structural formulae below and circle the ones with polar function groups. Remember polar functional groups always contain nitrogen (N), oxygen (O), phosphorus (P), and/or sulfur (S).

a.

$$H-\overset{\overset{\displaystyle H}{|}}{\underset{\underset{\displaystyle H}{|}}{C}}-\overset{\overset{\displaystyle H}{|}}{\underset{\underset{\displaystyle H}{|}}{C}}=\overset{\overset{\displaystyle H}{|}}{C}-\overset{\overset{\displaystyle H}{|}}{\underset{\underset{\displaystyle H}{|}}{C}}-SH$$

b.

$$H-\overset{\overset{\displaystyle H}{|}}{\underset{\underset{\displaystyle NH_2}{|}}{C}}-\overset{\overset{\displaystyle O}{\|}}{C}-\overset{\overset{\displaystyle H}{|}}{\underset{\underset{\displaystyle H}{|}}{C}}-H$$

c.

$$\overset{\displaystyle H}{\underset{\displaystyle H}{}}C=\overset{\overset{\displaystyle H}{|}}{\underset{\underset{\displaystyle H}{|}}{C}}-\overset{\overset{\displaystyle H}{|}}{\underset{\underset{\displaystyle H}{|}}{C}}-H$$

d.

$$H-\overset{\overset{\displaystyle H}{|}}{\underset{\underset{\displaystyle H}{|}}{C}}-\overset{\overset{\displaystyle H}{|}}{\underset{\underset{\displaystyle OH}{|}}{C}}-\overset{\overset{\displaystyle O}{\|}}{C}-OH$$

e.

$$H-\overset{\overset{\displaystyle H}{|}}{\underset{\underset{\displaystyle H}{|}}{C}}-\overset{\overset{\displaystyle H}{|}}{\underset{\underset{\displaystyle H}{|}}{C}}-\overset{\overset{\displaystyle H}{|}}{\underset{\underset{\displaystyle H}{|}}{C}}-\overset{\overset{\displaystyle H}{|}}{\underset{\underset{\displaystyle H}{|}}{C}}-H$$

f.

$$H-\overset{\overset{\displaystyle H}{|}}{\underset{\underset{\displaystyle NO_3}{|}}{C}}-\overset{\overset{\displaystyle NO_3}{|}}{\underset{\underset{\displaystyle H}{|}}{C}}-\overset{\overset{\displaystyle H}{|}}{\underset{\underset{\displaystyle NO_3}{|}}{C}}-H$$

g.

$$H-\overset{\overset{\displaystyle H}{|}}{\underset{\underset{\displaystyle H}{|}}{C}}-\overset{\overset{\displaystyle H}{|}}{\underset{\underset{\displaystyle CH_3}{|}}{C}}-\overset{\overset{\displaystyle H}{|}}{\underset{\underset{\displaystyle H}{|}}{C}}-\overset{\overset{\displaystyle O}{\|}}{C}-H$$

h.

$$H-\overset{\overset{\displaystyle H}{|}}{\underset{\underset{\displaystyle H}{|}}{C}}-\overset{\overset{\displaystyle H}{|}}{\underset{\underset{\displaystyle H}{|}}{C}}-\overset{\overset{\displaystyle H}{|}}{\underset{\underset{\displaystyle CH_3}{|}}{C}}-H$$

i.

$$O=\overset{\overset{\displaystyle HO}{}}{C}\diagdown \underset{\underset{\displaystyle H}{}}{C}=\overset{\overset{\displaystyle OH}{}}{\underset{\underset{\displaystyle H}{}}{C}}\diagup C=O$$

j.

$$\overset{\displaystyle H_3C}{}\diagdown \underset{\underset{\displaystyle H}{}}{C}-\overset{\overset{}{}}{\underset{\underset{\displaystyle H}{}}{C}}\diagup \overset{\displaystyle CH_3}{}$$

Notes:

Name: _______________________________

crn: _______________________________

Outcomes
Identify major organic functional groups and explain the properties of each.
Discuss how the properties of a functional group alter the chemical properties of an organic molecule.

A. Prelab Assignment

Alcohols are classified as primary (1°), secondary (2°), or tertiary (3°). A primary alcohol has one or no carbon atoms directly bonded to the alcohol carbon. The alcohol carbon is the carbon to which the hydroxyl group (-OH) is bonded. Examples of primary alcohols are methanol, ethanol, 1-propanol, and 1-butanol. The hydroxyl group could be bonded to either carbon in ethanol. In 1-propanol and 1-butanol, the hydroxyl group could be bonded to the left most carbon instead of the right end carbon.

Methanol	Ethanol	1-propanol	1-butanol
| ```
  H
  |
H-C-H
  |
  OH
``` | ```
  H  OH
  |  |
H-C--C-H
  |  |
  H  H
``` | ```
   H  H  H
   |  |  |
HO-C--C--C-H
   |  |  |
   H  H  H
``` | ```
  H  H  H  H
  |  |  |  |
H-C--C--C--C-OH
  |  |  |  |
  H  H  H  H
``` |

In a secondary alcohol, there are two carbons directly bonded to the alcohol carbon. Examples of secondary alcohols are 2-propanol, 2-butanol, 3-pentanol (below), and cyclohexanol (see p. 74).

2-propanol	2-butanol	3-pentanol
| ```
  H    H    H
  |    |    |
H-C----C----C-H
  |    |    |
  H    OH   H
``` | ```
  H    H    OH   H
  |    |    |    |
H-C----C----C----C-H
  |    |    |    |
  H    H    H    H
``` | ```
  H    H    H    H    H
  |    |    |    |    |
H-C----C----C----C----C-H
  |    |    |    |    |
  H    H    OH   H    H
``` |

Tertiary alcohols are alcohols in which there are three carbons directly bonded alcohol carbon. Examples tertiary alcohols are 2-methyl-2-propanol (isopropyl alcohol) and 2-methyl-2-butanol (*tert*-butanol).

2-methyl-2-propanol	2-methyl-2-butanol
| ```
        H
        |
      H-C-H
  H     |     H
  |     |     |
H-C-----C-----C-H
  |     |     |
  H     OH    H
``` | ```
  H     OH   H    H
  |     |    |    |
H-C-----C----C----C-H
  |     |    |    |
  H     |    H    H
      H-C-H
        |
        H
``` |
```

BIO1100
Lab 9 Alcohols

Study the structural formulae of the alcohols below and then mark each as a primary (1°), secondary (2°), or tertiary (3°) alcohol.

1.

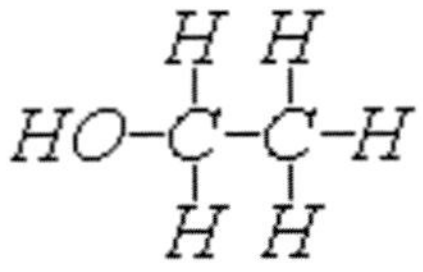

____

2.

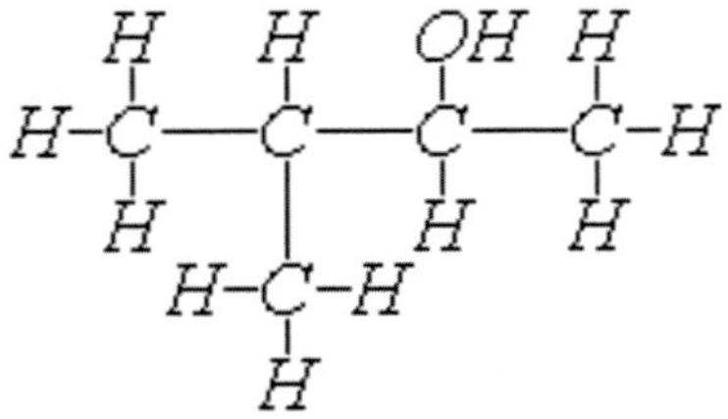

____

3.

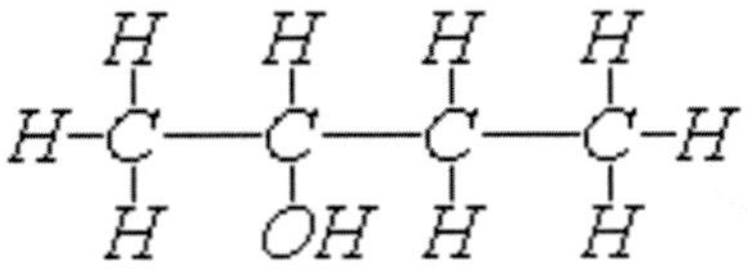

____

4.

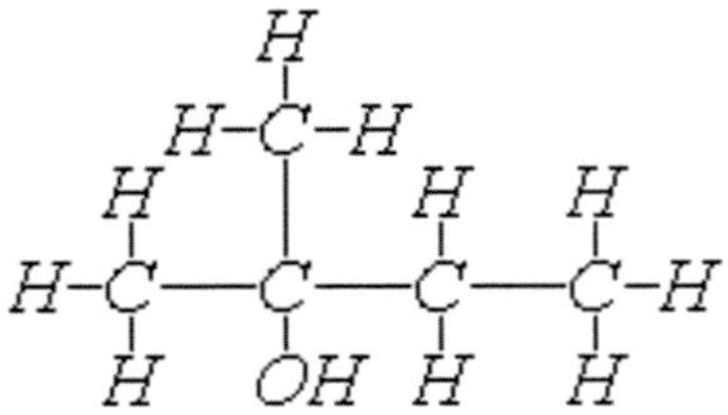

____

5.

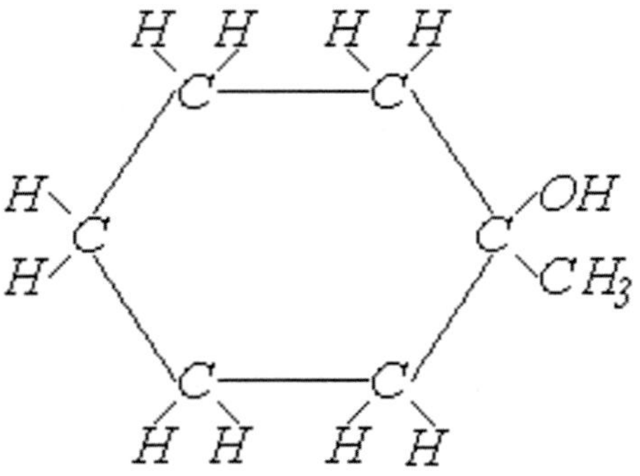

____

6.

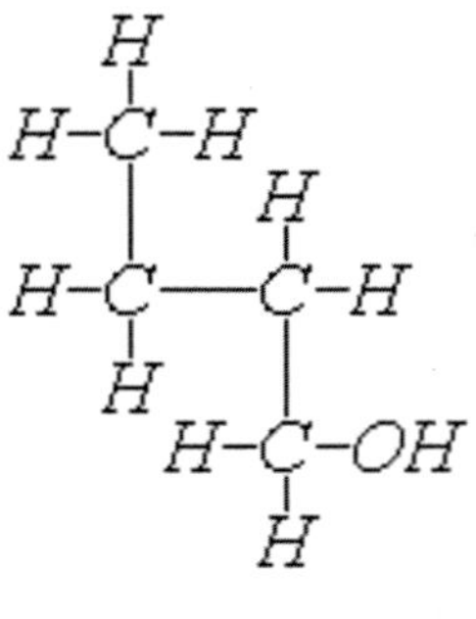

____

B. Alcohol Production

!!! SAFETY GOGGLES MUST BE WORN DURING METHANOL PRODUCTION!!!

Robert Boyle is credited with first producing methanol from the destructive distillation of boxwood (*Buxus sp.*) in 1661. For this reason methanol is commonly called wood alcohol to distinguish it from ethanol which is sometimes referred to as grain or drinking alcohol. Although animals can metabolize small amounts of ethanol, it would be a serious and perhaps fatal mistake to drink methanol.

1. Methanol Production

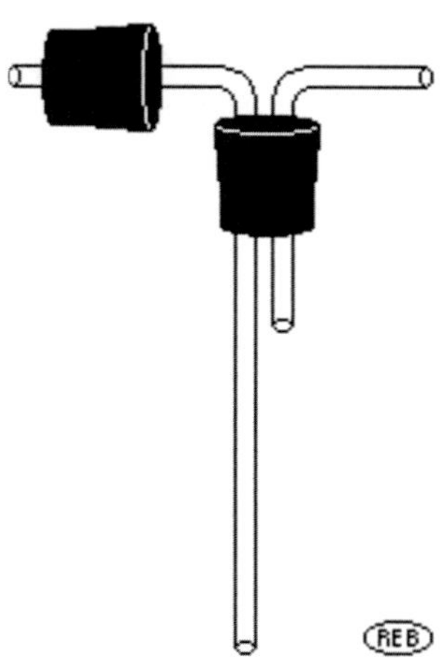

a. Carry a 250 mL beaker to the fume hood and set it down in the fume hood. Go to the supply bench and obtain two rubber stoppers with glass tubing as shown above and two large test tubes (25 x 150 mm). Carefully handle the rubber stoppers and glass tubing so you do not break the glass tubing. Carry the test tubes, rubber stoppers, and glass tubing back to the fume hood and place them in the 250 mL beaker.

b. Return to the supply bench and obtain a vertical support and a test tube clamp. Take these back to the fume hood. Set one of the test tubes in a 250 mL beaker. Clamp the other test tube with the test tube clamp at the opening of the test tube. Attach the test tube clamp with test tube to the vertical support. Attach the test tubes to the rubber stoppers so you have the set-up shown below.

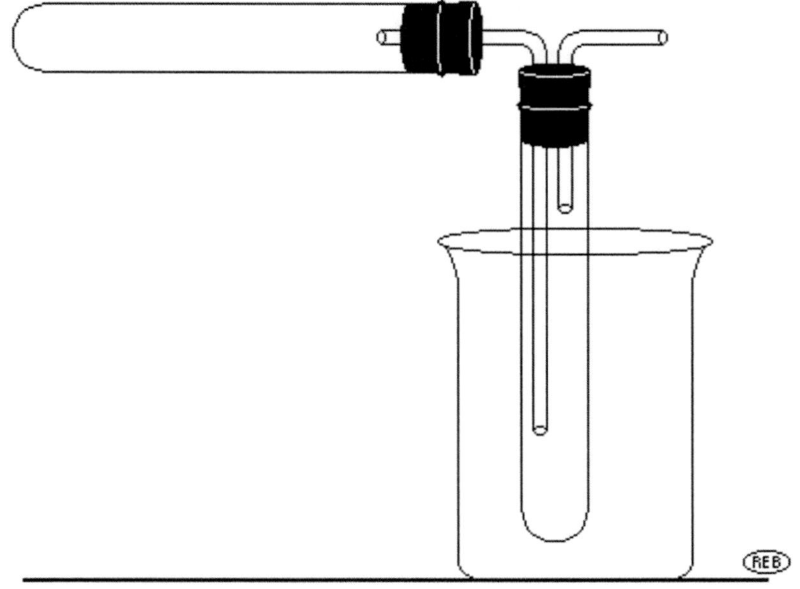

c. Use a flask or beaker and fill the beaker containing the test tube with water so that the beaker is about three-quarters full.

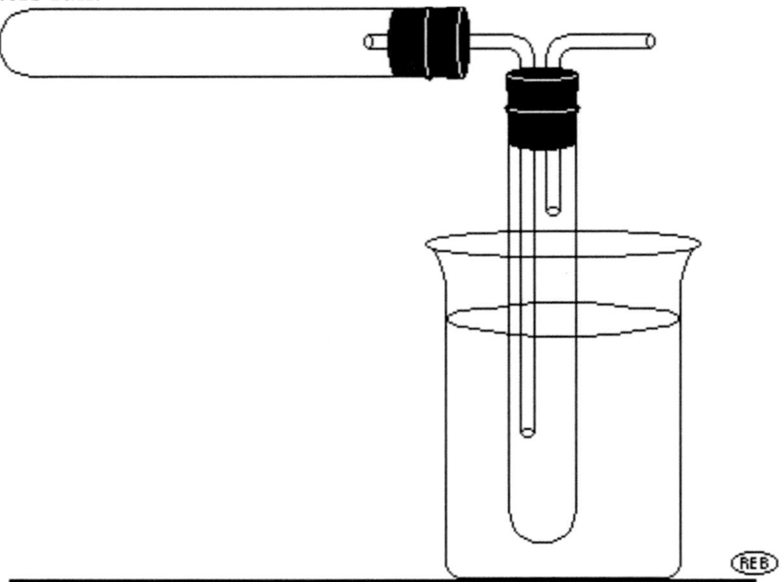

d. Loosen the clamp on the horizontal test tube and remove it from the rubber stopper. Take the test tube to the supply bench. Pack sawdust into test tube. Return the test tube to fume hood and reattach the test tube packed with sawdust to the rubber stopper and the test tube clamp.

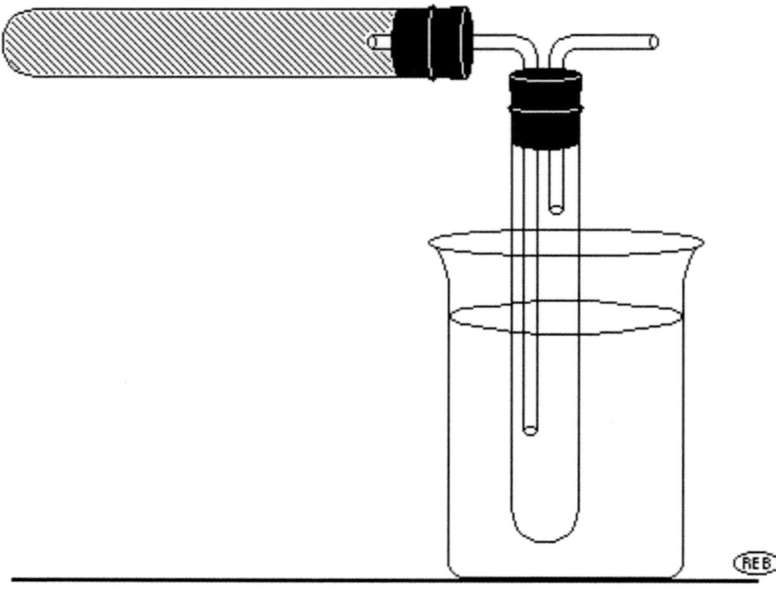

e. You should be wearing your safety goggles. Light the Bunsen burner and adjust the flame. Do not burn the rubber stopper. Pass the flame back and forth along the length of the test tube with saw dust. Do not burn the rubber stopper. As you heat the test tube, observe what is happening to the sawdust. Continue heating the test tube until you see drops of distillate collecting at the bottom of the sealed test tube.

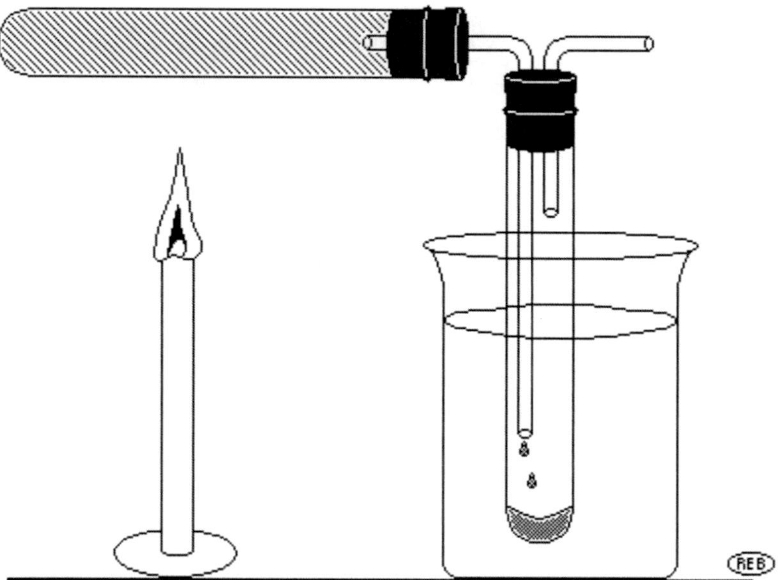

f. Continue to heat the test tube with sawdust until the distillate ceases. Do not burn the rubber stopper. Allow the distilling apparatus to cool. Remove the collection test tube with the distillate and set the test tube in a beaker. Do not pour the distillate into the beaker. Dissemble the remaining distilling apparatus and return the parts to where you found them on the lab supply bench. Take the beaker with the collection test tube and distillate to your lab bench.

g. Obtain a piece of #4 filter paper (110 mm) from the lab bench and take it to your lab bench. Place your funnel in your 125 mL Erlenmeyer flask. Fold the filter paper in half and then fold the again in half. Unfold the filter paper into the basin of your funnel. Spread out the filter paper and pour the distillate into the filter lined funnel.

h. Use a grease pencil to label a small test tube (16 x 120 mm) with your name and crn. Transfer your distillate into the test tube. Observe the color and smell of the distillate. Remember to use wafting technique when smelling the distillate. Record your observations below, then take your distillate in the labeled test tube to your instructor.

____

____

____

____

____

____

____

C. Alcohol Classification

!!! SAFETY GOGGLES MUST BE WORN DURING ALCOHOL CLASSIFICATION!!!

You were introduced to the structural differences of 1°, 2°, and 3° alcohols in your pre-lab assignment. You are now going to learn how to classify alcohols based on differences in their chemical reactivity. The differences in chemical reactivity are a result of their structural differences.

Oxidation is the removal of e^- from a substance. This can be accomplished by the addition of oxygen atoms and/or the removal of hydrogen atoms. Reduction is the addition of e^- to a substance. This can be accomplished by the removal of oxygen atoms and/or the addition of hydrogen. Notice that these two reactions are complementary. An oxidation is always coupled with a reduction. Primary alcohols can be oxidized to form aldehydes (controlled oxidation) which will further oxidize to form carboxylic acids (complete oxidation). Secondary alcohols are oxidized to form ketones. Tertiary alcohols do not oxidize. Potassium permanganate is shown as the oxidizer but many other oxidizers could be used. **When potassium permanganate is used as the oxidizer, a positive test is indicated by the KMnO4 becoming less purple and more reddish brown.**

For primary alcohols (by-products not shown)
Controlled Oxidation

$$\underset{1^o \text{ alcohol}}{R\text{-}\overset{\displaystyle OH}{\underset{\displaystyle H}{C}}\text{-}H} \quad + \quad \underset{\text{oxidizer}}{KMnO_4} \quad \longrightarrow \quad \underset{\text{aldehyde}}{R\text{-}\overset{\displaystyle O}{C}\text{-}H}$$

And then
Complete Oxidation

$$\underset{\text{aldehyde}}{R\text{-}\overset{\displaystyle O}{C}\text{-}H} \quad + \quad \underset{\text{oxidizer}}{KMnO_4} \quad \longrightarrow \quad \underset{\text{carboxylic acid}}{R\text{-}\overset{\displaystyle O}{C}\text{-}OH}$$

For secondary alcohols (by-products not shown)

$$\underset{2^o \text{ alcohol}}{R\text{-}\overset{\displaystyle OH}{\underset{\displaystyle H}{C}}\text{-}R} \quad + \quad \underset{\text{oxidizer}}{KMnO_4} \quad \longrightarrow \quad \underset{\text{ketone}}{R\text{-}\overset{\displaystyle O}{C}\text{-}R}$$

BIO1100
Lab 9 Alcohols

Third degree and second degree alcohols will substitute their hydroxyl group for a chloride group when treated with a solution of anhydrous zinc chloride in concentrated hydrochloric acid. The reaction is a nucleophilic substitution in which a chloride from HCl replaces the hydroxyl group. The anhydrous $ZnCl_2$ functions as a catalyst. Primary alcohols will not undergo this reaction. A solution of anhydrous $ZnCl_2$ in concentrated HCl is called Lucas' Reagent after Howard Lucas who developed the test. Tertiary alcohols react very quickly, usually within 10 minutes, Secondary alcohols react slowly and many times need to be heated for the reaction to go forward. **In either case, a positive result is indicated by white water-insoluble product, a chloroalkane, which causes the test solution to become turbid and cloudy.**

For secondary alcohols

$$\begin{array}{ccccccc}
 & R & & & & R & \\
 & | & & ZnCl_2 & & | & \\
H-C-OH & + \ HCl & \xrightarrow{\hspace{1.5cm}} & & H-C-Cl & + \ H_2O \\
 & | & & & & | & \\
 & R & & & & R & \\
2^o\ \text{alcohol} & & & & \text{chloroalkane} &
\end{array}$$

For tertiary alcohols

$$\begin{array}{ccccccc}
 & R & & & & R & \\
 & | & & ZnCl_2 & & | & \\
R-C-OH & + \ HCl & \xrightarrow{\hspace{1.5cm}} & & R-C-Cl & + \ H_2O \\
 & | & & & & | & \\
 & R & & & & R & \\
3^o\ \text{alcohol} & & & & \text{chloroalkane} &
\end{array}$$

1. KMnO$_4$ Oxidation test

a. Obtain a test tube rack from the supply bench. Mark five clean and dry 16 x 120 mm test tubes 1, 2, 3, D, and C. Place the test tubes in the test tube rack.

b. Take your test tube rack with test tubes to the supply bench and add 2 mL of $KMnO_4$ solution to each tube. One full transfer pipette is about 1 mL.

c. Take the test rack with test tubes to the fume hood. There are three unknown alcohols in the fume hood labeled 1, 2, and 3. Place 1 mL of each unknown into the appropriate test tube. Return to your lab bench. Place 1 mL of your crude methanol distillate into test tube D and 1 mL of RO water into test tube C. Test tube C is the control.

d. Gently swirl each test tube to mix the contents and let set while you start the Lucas tests.

e. An oxidation reaction will change the color of your test solution from purple to reddish-brown. Record any positive result with a + and any negative result with a − in the data table below.

2. Lucas test

a. Obtain a test tube rack from the supply bench. Mark five clean and dry 16 x 120 mm test tubes 1, 2, 3, D, and C. Place the test tubes in the test tube rack.

b. Take your test tube rack with test tubes to the supply bench and add 16-20 drops of Lucas' Reagent to each tube.

c. Take the test rack with test tubes to the fume hood. There are three unknown alcohols in the fume hood labeled 1, 2, and 3. Place 1 mL of each unknown into the appropriate test tube. Return to your lab bench. Place 1 mL of your crude methanol distillate into test tube D and 1 mL of RO water into test tube C. Test tube C is the control.

d. Mix well by placing a stopper in each tube and striking each tube vigorously with your finger. Be careful not to knock the test tube out of your hand. Observe for cloudiness.

e. After 10 minutes check for cloudiness again. Heat any test tubes that are clear in the dry water bath at 60 $^{\circ}$C for 15 minutes.

f. Record any tube that is cloudy as + and any tube that is clear as – in the data table below. Remember to return to $KMnO_4$ Oxidation test and interpret its results. Dispose of all test solutions in the sink.

3. **Data Table**- Based on your results classify each unknown and you distillate as a primary, secondary, or third degree alcohol.

Unknown	$KMnO_4$ (+ or -)	Lucas' Test (+ or -)	1°, 2°, or 3°
1			
2			
3			
D			
Control			

4. Oxidation of an alcohol always involves removal of ________________, which is accomplished by addition of ________________ atoms or removal of ________________ atoms.

5. Reduction of an alcohol always involves addition of ________________, which is accomplished by removal of ________________ atoms or addition of ________________ atoms.

6. Primary alcohols oxidize to form ________________, which then oxidize to form ________________.

7. Secondary alcohols oxidize to form ________________.

8. Tertiary alcohols oxidize to form ________________.

9. Return your lab tray to its cabinet.

Name: ___________________________________

crn: ___________________________________

Outcomes

Identify major organic functional groups and explain the properties of each.
Discuss how the properties of a functional group alter the chemical properties of an organic
  molecule.

A. Prelab Assignment

Use printed or internet sources to define the terms below.

1. Alcohol: ___________________________________

____

____

____

2. Carboxylic Acid: ___________________________________

____

____

____

3. Ester: ___________________________________

____

____

____

4. Triglyceride: ___________________________________

____

____

____

5. Ester bond: ___________________________________

____

____

____

B. Esterification

Esters occur widely in nature. Esters are organic compounds that contain ester bonds. Triglycerides, also called triacylglycerols or triacylglycerides, are long chained fatty acids ester bonded to glycerol (1, 2, 3-propanitriol). Cell membranes are composed of a phospholipid bilayer. Phospholipids are modified triglycerides. Waxes are long chain fatty acids ester bonded to long chain alcohols. Many of the fragrances and scents of plants are esters. Esters are produced by reacting an alcohol and a carboxylic acid with a sulfuric acid catalyst and an increased temperature. The reaction below shows a primary alcohol but a secondary or tertiary alcohol could also be used

$$R\text{-}\overset{\overset{H}{|}}{\underset{\underset{H}{|}}{C}}\text{-OH} \quad + \quad HO\text{-}\overset{\overset{O}{\|}}{C}\text{-R} \quad \xrightarrow[\uparrow T^\circ]{H_2SO_4} \quad R\text{-}\overset{\overset{H}{|}}{\underset{\underset{H}{|}}{C}}\text{-O-}\overset{\overset{O}{\|}}{C}\text{-R} \quad + \quad H_2O$$

      alcohol   carboxylic acid   ester

C. Naming Esters Compounds

Esters are named by combining the names of the reactant alcohol and carboxylic acid. The alcohol is named as an akyl and the carboxylic acid is named as its ion. In general terms, esters are alkyl carboxylates.

Example 1- What would be the name of an ester formed from the reaction of ethanol and methanoic acid (formic acid)?

    First-   Drop the -anol in ethanol and change it to -yl; ethyl.

    Next-   Drop the -ic acid in methanoic acid and change it to -ate; methanoate.

    Last-   Combine the two; ethyl methanoate.

$$H\text{-}\overset{\overset{H}{|}}{\underset{\underset{H}{|}}{C}}\text{-}\overset{\overset{H}{|}}{\underset{\underset{H}{|}}{C}}\text{-OH} \quad + \quad HO\text{-}\overset{\overset{O}{\|}}{C}\text{-H} \quad \xrightarrow[\uparrow T^\circ]{H_2SO_4} \quad H\text{-}\overset{\overset{H}{|}}{\underset{\underset{H}{|}}{C}}\text{-}\overset{\overset{H}{|}}{\underset{\underset{H}{|}}{C}}\text{-O-}\overset{\overset{O}{\|}}{C}\text{-H} \quad + \quad H_2O$$

  ethanol   methanoic acid   **ethyl methanoate**
                (formic acid)   **(ethyl formate)**

Example 2- What reactant alcohol and carboxylic acid would be needed to produce 1-propyl ethanote (1-propyl acetate)?

    First-   Drop the -yl in 1-propyl and change it to -anol; 1-propanol.

    Next-   Drop the -ate in ethanoate and change it to -ic acid; ethanoic acid.

    Last-   Write the alcohol and the carboxylic acid; 1-propanol and ethanoic acid

$$H\text{-}\overset{\overset{H}{|}}{\underset{\underset{H}{|}}{C}}\text{-}\overset{\overset{H}{|}}{\underset{\underset{H}{|}}{C}}\text{-}\overset{\overset{H}{|}}{\underset{\underset{H}{|}}{C}}\text{-OH} \quad + \quad HO\text{-}\overset{\overset{O}{\|}}{C}\text{-}\overset{\overset{H}{|}}{\underset{\underset{H}{|}}{C}}\text{-H} \quad \xrightarrow[\uparrow T^\circ]{H_2SO_4} \quad H\text{-}\overset{\overset{H}{|}}{\underset{\underset{H}{|}}{C}}\text{-}\overset{\overset{H}{|}}{\underset{\underset{H}{|}}{C}}\text{-}\overset{\overset{H}{|}}{\underset{\underset{H}{|}}{C}}\text{-O-}\overset{\overset{O}{\|}}{C}\text{-}\overset{\overset{H}{|}}{\underset{\underset{H}{|}}{C}}\text{-H} \quad + \quad H_2O$$

  1-propanol   **ethanoic acid**   1-propyl ethanoate
                 (acetic acid)   (1-propyl acetate)

BIO1100
Lab 10 Esters

Exercise 1- Given the reactant alcohol and carboxylic acid, name and write the full structural formula of the ester product and write the molecular formula of the by-product.

a. Ethanol (ethyl alcohol) and butanoic acid (butyeric acid). This ester is in the fragrance of pineapple.

```
  H H                    O H H H
  | |                    || | | |          H2SO4
H-C-C-OH      +      HO-C-C-C-C-H       ───────────→
  | |                       | | |            ↑T°
  H H                       H H H

   ethanol              butanoic acid
(ethyl alcohol)          (butyric acid)
```

b. 1-octanol and ethanoic acid (acetic acid). This ester is in the fragrance of oranges.

```
  H H H H H H H H                    O H
  | | | | | | | |                    || |          H2SO4
H-C-C-C-C-C-C-C-C-OH      +      HO-C-C-H       ───────────→
  | | | | | | | |                       |            ↑T°
  H H H H H H H H                       H

     1-octanol                     ethanoic acid
                                   (acetic acid)
```

c. Methanol and salicylic acid. This ester is in the fragrance of wintergreen.

```
     H                           O
     |                           ||             H2SO4
   H-C-OH        +        HO-C─⟨ ⟩          ───────────→
     |                       HO─⟨ ⟩             ↑T°
     H

   methanol                salicylic acid
 (formyl alcohol)
```

Exercise 2- Given the product and by-product of an ester reaction, name and write the full structural formula of the reactant alcohol and carboxylic acid.

a. Methyl-*trans*-cinnamate. This ester is in the fragrance of strawberries.

```
                              H  O  H
    H2SO4                     |  ||  |
 ───────────→            H-C-O-C-C=C-⟨ ⟩        +      H2O
    ↑T°                      |      |
                            H      H

                       methyl-trans-cinnamate
```

b. 3-methyl-1-butyl ethanoate (3-methyl-1-acetate). This ester is in the fragrance of bananas.

```
                                      H
                                      |
                                    H-C-H
    H2SO4                   H  |  H H    O H
 ───────────→              |  |  | |    || |           +      H2O
    ↑T°                  H-C-C-C-C-O-C-C-H
                           |  |  | |       |
                           H  H  H H       H

                      3-methyl-1-butyl ethanoate
```

C. Ester Synthesis

!!! SAFETY GOGGLES MUST BE WORN DURING ALL ESTER REACTIONS!!!

1. Place a clean, dry test tube into a test tube rack. Carry the test tube rack and test tube to the lab supply bench. Use a disposable pipette to transfer 2 mL of concentrated ethanoic acid (acetic acid) into the test tube. Remember that 2 mL is about two full transfer pipettes.

2. Next, add 2 mL of 1-pentanol (n-amyl alcohol) to the same test tube. Gently swirl the mixture in the test tube.

3. Tilt the test tube to an angle of approximately 45° and carefully add 6-8 drops of concentrated H_2SO_4 to the test tube. Use a clean stirring rod and gently stir the mixture.

4. Carry the test tube rack with the test tube and the reacting mixture to a dry water bath. The dry water should be set to $60^{\circ}C$. Place the test tube in the dry water bath for 20 minutes.

5. Remove the test tube from the dry water bath and all the test tube to cool. When the test tube is cool, pick it up and waft its fumes. Describe the smells you are sensing as pleasant or unpleasant, aromatic or acrid, sweet or sour, and woody or fruity. After you have described the contents of your ester reaction, dispose of contents in the sink.

---

6. Show the complete chemical reaction for the esterification reaction you just performed. Write the full structural formulae for the reactants and products. Include any catalysts or special conditions required for the reaction. Write the names of the reactants and products under their structural formulae and circle the ester bond.

7. Return your lab tray to its cabinet.

Notes:

Name: _______________________________

crn: _______________________________

Outcomes

Describe the cellular functions and physical properties of carbohydrates, lipids, proteins, and nucleic acids.

A. Prelab Assignment

Use printed or internet sources to define the terms below.

1. Biochemistry: _______________________________

____

____

____

2. Monosaccharide: _______________________________

____

____

____

3. Fatty acid _______________________________

____

____

____

4. Amino Acid: _______________________________

____

____

____

5. Nucleotide: _______________________________

____

____

____

Lab 11 Biochemistry Models

B. Biochemical Models

!!! SAFETY GOGGLES ARE NOT REQUIRED FOR BIOCHEMICAL MODELS!!!

You are going to construct some three dimension models of representative molecules of biochemical compounds. Obtain a molecular model kit from the lab supply bench. Take the kit back to your lab bench and open it up. Remember, black spheres represent carbon atoms and will simply be referred to as carbon atoms or carbons. Hydrogen atoms are generally represented by yellow or white spheres while oxygen is usually a red sphere and nitrogen a light blue sphere. Notice that there are three kinds of connectors representing covalent bonds. The short connectors are used to connect hydrogen atoms to carbon and other larger atoms such as nitrogen and oxygen. While the long connectors are used to connect carbon to carbon and carbon to nitrogen and oxygen. The spring connectors are used to form double and triple covalent bonds.

1. Glucose (Dextrose), $C_6H_{12}O_6$

   a. Below are two dimensional representations of glucose. Glucose is the primary energy source for eukaryotic cells. Glucose exists in two general forms: the straight chain form and the closed ring form. The straight chain form exists as either D (right handed) or L (left handed) stereoisomers. The closed ring form, in addition to either being D or L can also be either α (alpha) or β (beta) anomers.

Glucose	
Fisher	Haworth
D-glucose	α-D-glucose

   b. Construct a Fisher model of the straight chain form of glucose. Remember spring connectors are used to form double bonds like the double bound present in the aldehyde group between oxygen and the first carbon. When the aldehyde group is oriented at the top carbon chain. The top carbon is designated the first carbon. The hydroxyls on the carbon chain should match the hydroxyls in the drawing. Since only the fourth carbon has a hydroxyl on the left in the drawing, only the fourth carbon in your model should have a hydroxyl on the left.

   c. Take your Fisher model of D-glucose to your instructor.

   d. Change your Fisher model of D-glucose to a Haworth model of α-D-glucose. Separate the oxygen in the aldehyde group from the first carbon. Remove the two spring connectors from the carbon and oxygen. Remove the hydroxyl group (-OH) with its short wooden connector from the fifth carbon. Replace the hydroxyl group and its short connector with long wooden connector.

   e. Connect the other end of the long connector that you just placed in the fifth carbon to the oxygen. Place another long connector into the second hole of the oxygen. Connect the other end of that long connector to the first carbon.

f. Place the short connector and hydroxyl group (-OH) that you removed from the fifth carbon into the open hole of the first carbon.

g. Take your Fisher model of α-D-glucose to your instructor.

2. Maleic acid (2-*cis*-butenedioic acid), $C_4H_4O_4$

a. Below are two dimensional representations of maleic acid and fumaric acid. Both these acids play a role in the Krebs cycle, also called the citric acid cycle (CAC) or the tricarboxylic acid (TCA) cycle. Note that both maleic acid and fumaric acid are dicarboxylic acids. That is they have a carboxyl function group at each end. The carboxyl functional group is an oxygen double bonded to a carbon with a hydroxyl group attached to the carbon.

2- Butenedioic acid isomers	
$$HO-\overset{\overset{O}{\|\|}}{C}-\overset{\underset{H}{\|}}{C}=\overset{\underset{H}{\|}}{C}-\overset{\overset{O}{\|\|}}{C}-OH$$	$$HO-\overset{\overset{O}{\|\|}}{C}-\overset{\underset{H}{\|}}{C}=\overset{\overset{H}{\|}}{C}-\overset{\overset{O}{\|\|}}{C}-OH$$
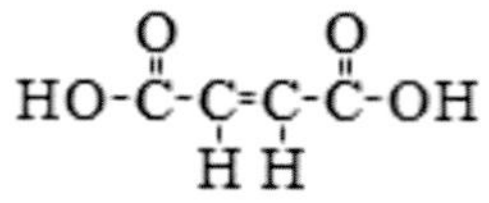 **Maleic acid** **(2-*cis*-butenedioic acid)**	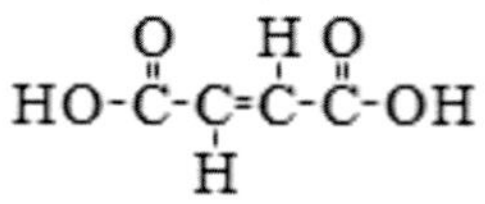 **Fumaric acid** **(2-*trans*-butenedioic acid)**

b. Study the full structural formula of maleic acid, and then construct a three dimensional model of it.

c. Take your model of maleic acid to your instructor.

d. Study the full structural formula of fumaric acid, construct a three dimensional model of it, and then take your model of fumaric acid to your instructor.

3. *cis*-oleic acid, $C_{18}H_{34}O_2$

a. *cis*-Oleic acid is a common fatty acid found in nature. It is especially present in olives from whence it gets its name. Oleic acid an important constituent of the phospholipids that comprise cell membranes. Oleic acid is a long chain fatty acid with a double bond between the ninth and tenth carbon and a single carboxyl group at one end of the carbon chain.

b. Construct a ball and stick model of *cis*-oleic acid and take it to your instructor.

4. L-alanine, $C_3H_7NO_2$

a. All proteins found in nature contain L- (left-handed) amino acids. Amino acids have a central carbon with an amino group, a carboxylic acid group, a hydrogen, and an R group. The simplest amino acid is glycine in which the R group is hydrogen. The variations in the R groups give the individual amino acids their characteristics. In the drawing below imagine the hydrogen atom and the methyl group on the central carbon coming out of the paper at you and the amino and carboxyl group retreating into the paper.

<table>
<tr><td colspan="2" align="center">Alanine Enatiomers</td></tr>
<tr>
<td align="center">

H H O
H-N-C-C-OH
H-C-H
H

</td>
<td align="center">

O H H
HO-C-C-N-H
H-C-H
H

</td>
</tr>
<tr>
<td align="center">L-alanine
(L-2-aminopropanoic acid)</td>
<td align="center">D-alanine
(D-2-aminopropanoic acid)</td>
</tr>
</table>

b. Place four long wooden connectors into a carbon atom. This will be your central carbon.

c. Orient your central carbon so that the vertical connectors are coming out at you and the horizontal connectors are going away from you. This will be called L orientation. Attach a hydrogen atom to the top vertical connector and attach a carbon atom to the bottom wooden connector. Place short wooden connectors into three hydrogen atoms

d. Hold the central carbon atom with the hydrogen atom up and the methyl group down. The two empty connectors should be horizontal. Place a nitrogen atom (light blue) on the left wooden connector and a carbon atom on the right wooden connector.

e. Place two spring connectors into an oxygen and then attach two spring connectors with the oxygen to the horizontal carbon.

f. Place a short connector into an oxygen and attach a hydrogen to that short connector. This red ball and yellow ball with the short connector represent a hydroxyl group (-OH). Use a short connector to attach this hydroxyl group to the horizontal carbon with the double bond oxygen.

g. Take your model of L-glycine to your instructor.

5. L-asparagine, $C_4H_8N_2O_3$

a. L-asparagine is the first amino acid to be isolated. As its name suggests asparagine was first isolated from asparagus. While asparagine is required for the development and function of the brain, asparagine is not an essential amino acid and can be synthesized from oxaloacetic acid.

<table>
<tr><td colspan="2" align="center">Asparagine and Oxaloacetic Acid</td></tr>
<tr>
<td align="center">

```
    H H O
    | | ‖
H-N-C-C-OH
    |
  H-C-H
    |
    C=O
    |
  H-N-H
```

L-asparagine

</td>
<td align="center">

```
      O H O O
      ‖ | ‖ ‖
HO-C-C-C-C-OH
      |
      H
```

Oxaloacetic acid

</td>
</tr>
</table>

b. Place four long wooden connectors into a carbon atom. This will be your central carbon.

c. Orient your central carbon to the L orientation. Attach an amino group to the left wooden connector of the central carbon and a carboxyl group to the right wooden connector of the central carbon.

d. With the central carbon in L orientation, attach a hydrogen to the top wooden connector.

e. Connect two carbons together using a long wooden connector. Connect a nitrogen to one of the carbons using a long wooden connector.

f. Attach two hydrogen atoms to the nitrogen using short wooden connectors. Attach two spring connectors to the carbon directly attached to the nitrogen. Attach an oxygen to the two spring connectors.

g. Use short connectors to attach hydrogens to the end carbon. Place a long wooden connector in the last hole of the end carbon. This is the R group of asparagine. Attach the R group to the central carbon.

h. Take your model to your instructor,

6. Adenine, $C_5H_5N_5$

a. Adenine is one of five nitrogen bases that occur in living things. The other four nitrogen bases are cytosine, guanine, thymine, and uracil. Adenine, cytosine, guanine, and thymine are the nitrogen bases found in deoxyribonucleic acid (DNA) nucleotides. Ribonucleic acid (RNA) nucleotides contain adenine, cytosine, guanine, and uracil. Adenosine triphosphate is single RNA nucleotide that provides energy for all cells.

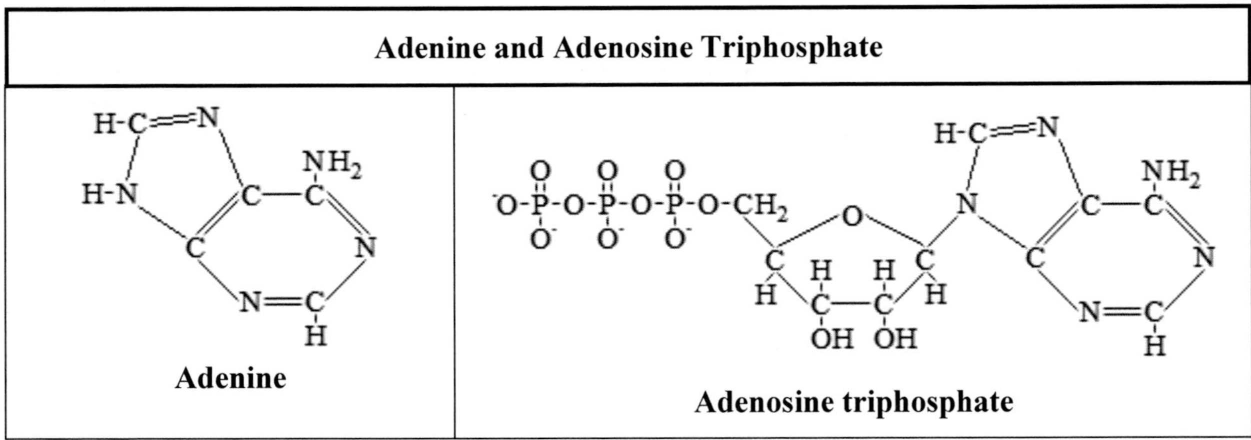

b. Study the full structural formula of adenine, and then construct a three dimensional model of it.

c. Take your model of adenine to your instructor.

Notes:

Name: ______________________________________

crn: ______________________________________

Outcomes
Describe the cellular functions and physical properties of carbohydrates, lipids, proteins, and nucleic acids.

A. Prelab Assignment

1. Carbohydrate: ______________________________________

____

____

____

2. Lipid: ______________________________________

____

____

____

3. Protein: ______________________________________

____

____

____

4. Nucleic acid: ______________________________________

____

____

____

B. Biochemical Tests

!!! SAFETY GOGGLES MUST BE WORN DURING ALL BIOCHEMICAL TESTS!!!

1. Benedict's Test

    a. Obtain a test tube rack from the supply bench. Label five clean test tubes (16 x 120 mm) 1, 2, 3, 4, and 5. Use a transfer pipette and carefully transfer a dropper full of glucose to test tube 1. Ensure that the amylose is properly suspended and use a transfer pipette to add a dropper full amylose to test tube 2. Use a transfer pipette to transfer a dropper full of albumin to test tube 3 and a dropper full of corn oil to test tube 4. Take the test tube rack with test tubes back to your lab bench and add one dropper full of RO water to the test tube marked 5. Test tube 5 is your control.

    b. Take your test tube rack with the five test tubes to the supply bench. Using a transfer pipette carefully add 8-10 drops of Benedict's solution to each tube.   Place your five test tubes into the dry water bath for 20 minutes. You will interpret and record the results in the data table below after you perform the Lugol's and Biuret test.

    c. A positive Benedict's test will range from greenish to reddish orange. A positive Benedict's is shown below. The test tube on the left is the positive the rest are negative.

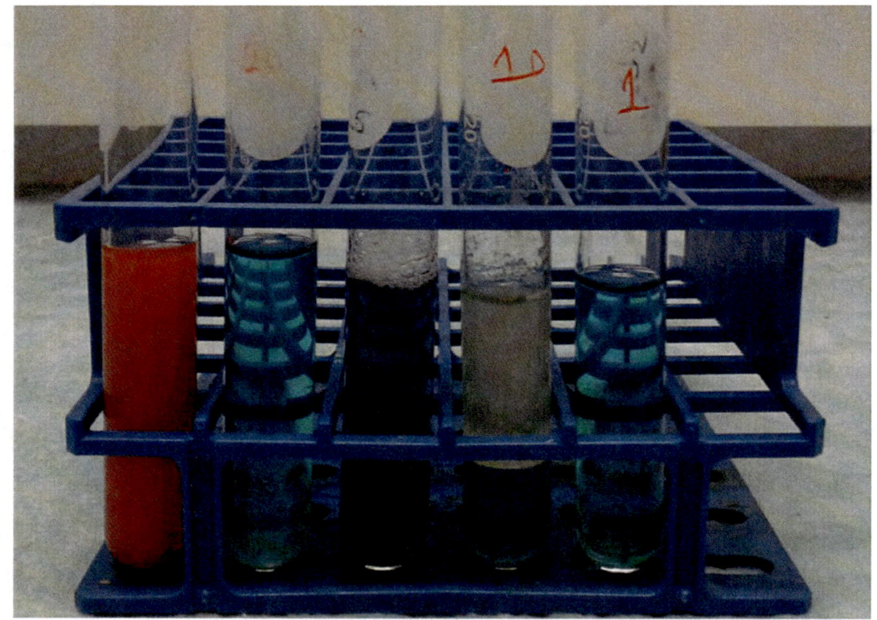

    d. All test solutions can be poured down the drain. Clean your test tubes with soap and brush. Rinse test tubes thoroughly.

2. Lugol's Test

a. Label five clean test tubes (16 x 120 mm) 1, 2, 3, 4, and 5. Use a transfer pipette and carefully transfer a dropper full of glucose to test tube 1. Ensure that the amylose is properly suspended and use a transfer pipette to add a dropper full amylose to test tube 2. Use a transfer pipette to transfer a dropper full of albumin to test tube 3 and a dropper full of corn oil to test tube 4. Take the test tube rack with test tubes back to your lab bench and add one dropper full of RO water to the test tube marked 5. Test tube 5 is your control.

b. Take your test tube rack with the five test tubes to the supply bench. Using a transfer pipette carefully add 2-4 drops of Lugol's solution to each tube.

c. A positive Lugol's test will result in a color change to blueish black. A positive Lugol's is shown below. The test tube second from the left is the positive the rest are negative. Record your results in the data table below.

d. All test solutions disposed of in the iodine waste bottle which are in the fume hood. Clean your test tubes with soap and brush. Rinse test tubes thoroughly.

3. Biuret Test

a. Label five clean test tubes (16 x 120 mm) 1, 2, 3, 4, and 5. Use a transfer pipette and carefully transfer a dropper full of glucose to test tube 1. Ensure that the amylose is properly suspended and use a transfer pipette to add a dropper full amylose to test tube 2. Use a transfer pipette to transfer a dropper full of albumin to test tube 3 and a dropper full of corn oil to test tube 4. Take the test tube rack with test tubes back to your lab bench and add one dropper full of RO water to the test tube marked 5. Test tube 5 is your control.

b. Take your test tube rack with the five test tubes to the supply bench. Using a transfer pipette carefully add 2-4 drops of Biuret solution to each tube.

c. A positive Biuret test will result in a color change to light to dark purple. A positive Biuret is shown below. The test tube in the middle is the positive the rest are negative. Record your results in the data table below.

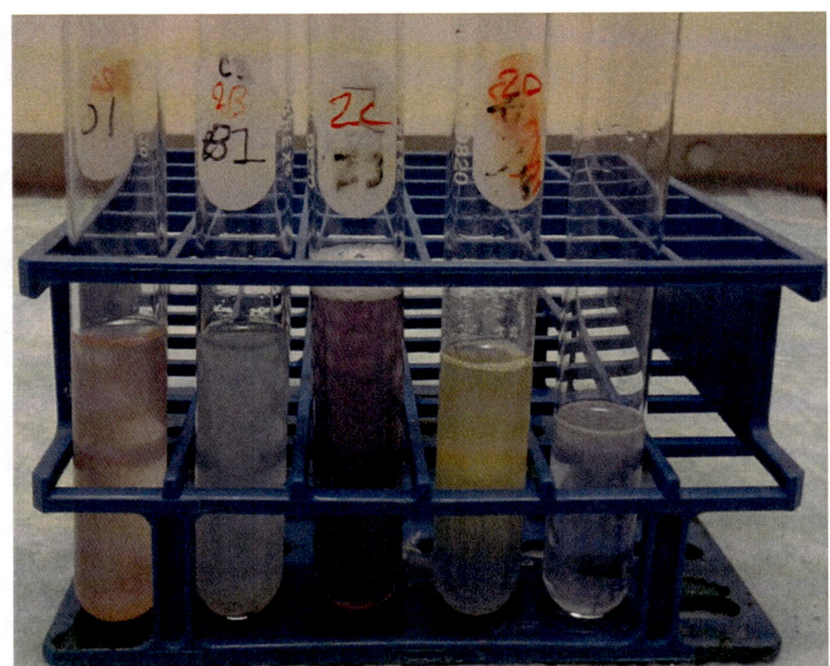

d. All test solutions can be poured down the drain. Clean your test tubes with soap and brush. Rinse test tubes thoroughly.

4. Solubility Test

a. Label five clean test tubes (16 x 120 mm) 1, 2, 3, 4, and 5. Use a transfer pipette and carefully transfer a dropper full of glucose to test tube 1. Ensure that the amylose is properly suspended and use a transfer pipette to add a dropper full amylose to test tube 2. Use a transfer pipette to transfer a dropper full of albumin to test tube 3 and a dropper full of corn oil to test tube 4. Take the test tube rack with test tubes back to your lab bench and add one dropper full of RO water to the test tube marked 5. Test tube 5 is your control.

b. Take your test tube rack with the five test tubes to the supply bench. Add about one mL of RO water to each test tube. One mL of RO water would be about the same volume that is already in the test tube. Gently swirl each test tube and use your stirring rod to thoroughly mix each solution. Remember to rinse and clean your stirring rod between solutions.

c. A positive Solubility test will result in the test solution uniformly mixing with water. A negative solubility test occurs when the test solution fails to mix with the water and separates forming two layers.

d. All test solutions can be poured down the drain. Clean your test tubes with soap and brush. Rinse test tubes thoroughly.

5. Data Table

a. Record your results as + (plus sign), meaning a positive result or − (minus sign) negative result. Unless your water control separated into two distinct layers, it should be recorded as a positive result with a plus sign.

Biochemical Tests					
Biochemical Class	**Specific Compound**	**Benedict's Test**	**Lugol's Test**	**Biuret Test**	**Solubility Test**
Monosaccharide	**Glucose, 1**				
Polysaccharide	**Amylose, 2**				
Protein	**Albumin, 3**				
Lipid	**Corn Oil, 4**				
Control, 5					

6. Analysis

a. Which test can be used to identify monosaccharides? ____

b. Which test can be used to identify polysaccharides? ____

c. Which test can be used to identify proteins? ____

d. Which test can be used to identify lipids? ____

C. Biochemical Unknowns

!!! SAFETY GOGGLES MUST BE WORN DURING ALL BIOCHEMICAL TESTS!!!

1. Unknowns

a. Obtain a test tube rack from the supply bench. Label three clean test tubes (16 x 120 mm) A, B, and C. Take your test tube rack with test tubes to the supply bench. Use a transfer pipette and carefully transfer a dropper full of Unknown A to test tube A and a dropper full of Unknown B to test tube B. Take the test tube rack with test tubes back to your lab bench and add a dropper full of RO water to the test tube marked C. Test tube C is your control.

b. Take your test tube rack with the three test tubes to the supply bench. Perform the four biochemical tests you have just learned to Unknown A, Unknown B, and the control. Record your results in the data table below and then identify the unknowns by answering the analysis questions.

2. Data Table

a. Record your results as + (plus sign), meaning a positive result or − (minus sign) negative result. Unless your water control separated into two distinct layers, it should be recorded as a positive result with a plus sign.

	Benedict's Test	Lugol's Test	Biuret Test	Solubility Test
Unknown A				
Unknown B				
Control				

3. Analysis

a. What biochemical compounds are present in Unknown A? ___________________________________

____

b. What biochemical compounds are present in Unknown B? ___________________________________

____

4. Return your lab tray to its cabinet.

Notes:

Name: _______________________________

crn: _______________________________

Outcomes
  Identify the major classes of lipids.
  Discuss the nature by which lipids form cellular membranes.

A. Prelab Assignment

Lipids are biologically important compounds. Phospholipids are an integral part of the cell membrane. The hydrophobic nature of lipids helps organisms prevent water lose. The hydrophobic nature of lipids is a result of their nonpolar hydrocarbon tails or rings. Lipids, because they are nonpolar, will not dissolve in water, a polar solvent. The main classes of biologically important lipids are fatty acids, waxes, triglycerides, steroids, phospholipids, and glycolipids. Fatty acids are long chain carboxylic acids. Most common biologically active fatty acids are made up of 12-18 carbons. Fatty acids are further divided into saturated, monounsaturated, and polyunsaturated fatty acids. Saturation refers to the number of hydrogen atoms attached to the carbon chain making up the fatty acid. A fatty acid that is unsaturated has at least one double carbon to carbon bond in the carbon chain. A monounsaturated fatty acid has only one carbon to carbon double bond in the carbon chain, while a polyunsaturated fatty acid has two or more carbon to carbon double bonds in the carbon chain. Draw the full structural formula for each following fatty acids and lipids compounds. Under each drawing write a sentence describing the biological and physiological importance of the compound.

Linoleic acid

Importance of linoleic acid? ____

____

Hyaluronic acid

Importance of hyaluronic acid? ____

____

Phospholipids

Importance of phopholipids? ____

____

B. Saturation

!!! SAFETY GOGGLES MUST BE WORN DURING ALL SATURATION TESTING!!!

Saturation in organic compounds describes the amount of hydrogen atoms bonded to the carbon atoms. Saturated carbons contain only hydrogen atoms. When an organic compound contains double or triple covalent bonds, the compound is unsaturated. The greater number of double and/or triple bonds the greater the unsaturation. Saturated fatty acids (SFA) and triglycerides contain no double or triple covalent bonds. Monounsaturated fatty acids (MFA or MUFA) contain at least one double covalent bond in their hydrocarbon tail. Polyunsaturated fatty acids (PFA or PUFA) contain two or more double covalent bonds in their hydrocarbon tail. Acetylenic fatty acids are PFA that contain triple covalent bonds.

1. Hanus Iodine Test

The alkene nature of unsaturated fatty allows them to be halogenated. Hanus iodine solution halogenates unsaturated fatty acids by adding I_2 across any double bonds that are present. The more double bonds present in a fatty acid means more I_2 can be added to the fatty acid. Hanus iodine is dark orange in color. So the reactant mixture starts out dark orange but as I_2 is added to any double bonds, the product mixture becomes lighter. If the product mixture remains dark in color then no double bonds are present to accept the I_2. This would be the case for a saturated fatty acid or triglyceride.

Hanus Iodine Test (by-products not shown)

$$\underset{\substack{\text{unsaturated}\\\text{fatty acid}}}{HO\text{-}\overset{\overset{O}{\|}}{C}\text{-}(CH_2)_{\overline{n}}\overset{H}{\underset{|}{C}}\text{=}\overset{H}{\underset{|}{C}}\text{-}(CH_2)_{\overline{n}}CH_3} \;+\; \underset{\text{iodine}}{I_2} \longrightarrow \underset{\substack{\text{halogenated}\\\text{fatty acid}}}{HO\text{-}\overset{\overset{O}{\|}}{C}\text{-}(CH_2)_{\overline{n}}\overset{I}{\underset{|}{C}}\text{=}\overset{I}{\underset{|}{C}}\text{-}(CH_2)_{\overline{n}}CH_3}$$

a. Obtain a test tube rack from the supply bench. Label four clean test tubes (16 x 120 mm) 1E, 2E, 3E, and 4E. These will be your experimental test tubes. 1E will receive linseed oil, 2E will receive corn oil, 3E will receive lard, and 4E will receive steric acid.

b. Add a finger width of each lipid to its appropriate test tube and place all four test tubes in a beaker half filled with water.

c. Place the beaker with all four test tubes on an electric heater to melt the lard and stearic acid.

d. Label four more test tubes (16 x 120 mm) 1C, 2C, 3C, and 4C. Add the same amount of lipid as before to the appropriate test tubes. These will be your controls.

e. Place this beaker with the controls on a separate electric heater to melt the lard and steric acid.

f. Remove the first four test tubes from the beaker when the lard and stearic acid have melted and add place them in a test tube rack. Remember, these are your experimental tubes.

g. Take the test tube rack with your experimental tubes to the fume hood and add 8 drops of the Hanus Iodine solution to each test tube.

h. Return the experimental test tubes to the beaker on the heater.

i. After 5 minutes, rank the experimental test tubes from darkest to lightest as compared to the control. In the drawing below, the experimental test tubes are in the first row. The control for each experimental test tube is behind it and to the right. The tubes have been ranked from left to right with the far right experimental test tube being the lightest compared to its control and, therefore the most reactive. While the far left test tube is the darkest compared to its control and the least reactive.

j. The experimental test tube that is darkest as compared to its control is the least reactive and should be ranked as- 4, darkest. Record your result in the data table below.

k. The experimental test tube that is lightest as compared to its control is the most reactive and should be ranked as- 1, lightest. Record your result in the data table below.

l. The remaining two test tubes should be ranked between 1 and 4, as 2, medium light and 3, medium dark. Record your result in the data table below.

m. All experimental solutions must be disposed of in the Iodine waste bottle. The contents of the control test tubes can poured down the sink. Flush the sink with hot water for at least two minutes. Clean your test tubes with soap and brush. Rinse test tubes thoroughly.

2. Data Collection & Analysis

a. Record your results in the table as directed. Iodine number is the grams of iodine that can be added to a 100 grams of a particular lipid.

Lipid	Iodine number	Darkness relative to control	Order of reactivity
Corn oil	116-130		
Lard	0		
Linseed oil	46-66		
Steric acid	175-205		

b. Do your experimental results agree with given iodine numbers? ________________________________

C. Emulsification

!!! SAFETY GOGGLES MUST BE WORN DURING ALL EMULSIFICATION TESTING!!!

Polar substances and non-polar substances do not mix evenly and permanently with each other. Immiscible describes substances that do not mix evenly with one another. Polar solvents dissolve polar solutes and non-polar solvents dissolve non-polar solutes. Emulsification involves mixing two immiscible substances with a third substance to allow the first two substances to mix evenly and permanently with each other. The third substance is called an emulsifying agent or a bridge solvent. You are going to test the relative effectiveness of two emulsifying agents to mix water and corn oil.

1. **Water-** You are familiar with the fact that water and corn oil are immiscible, that is to say they do not mix evenly and permanently with each other. Water is very polar and corn oil is very non-polar.

   a. Use a grease pencil and mark a clean test tube (16 x 120 mm) with a C.

   b. Use a graduated cylinder to add 5.0 mL of RO water to the test tube.

   c. Use a transfer pipette and add 1 dropper full of corn oil to the test tube

   d. Place a clean rubber stopper in the test tube and vigorously shake the test tube

   e. Remove the stopper and set the test tube in a test tube rack. This is your control.

2. **Ethanol-** Medieval apothecaries regularly used ethanol as a bridge solvent to mix the essential oils of medially beneficial herbs with water. While one carbon of ethanol is surrounded by hydrogen atoms and is very nonpolar, the other carbon has a hydroxyl group. The hydroxyl group gives this carbon substantial polar area. The polar area of ethanol allows it to readily dissolve in water. The nonpolar area of ethanol allows it to act as a solvent for oils and other nonpolar solutes.

   Ethanol

   $$H-\overset{\overset{\displaystyle H}{|}}{C}-\overset{\overset{\displaystyle H}{|}}{C}-OH$$

   a. Use a grease pencil and mark a clean test tube (16 x 120 mm) with an E.

   b. Use a graduated cylinder to add 5.0 mL of $1.0\%_{v/v}$ ethanol solution to the test tube.

   c. Use a transfer pipette and add 1 dropper full of corn oil to the test tube

   d. Place a clean rubber stopper in the test tube and vigorously shake the test tube

   e. Remove the stopper and set the test tube in a test tube rack.

3. Bile Acids- Bile is produced in the liver and stored in the gall bladder. Bile acids are the main component of bile. Bile acids aid in the digestion of lipids. Cholic acid is a typical bile acid. Like all bile acids, cholic acid is composed of three cyclohexane hydrocarbon rings and one cyclopentane hydrocarbon ring. These four hydrocarbon rings are the main nonpolar area of cholic acid and all bile acids. The carboxyl group and the three hydroxyl groups that are attached to cholic acid give it some polarity. So while the hydrocarbon rings can dissolve into corn oil, the polar hydroxyl and carboxyl groups can dissolve into water.

Cholic Acid

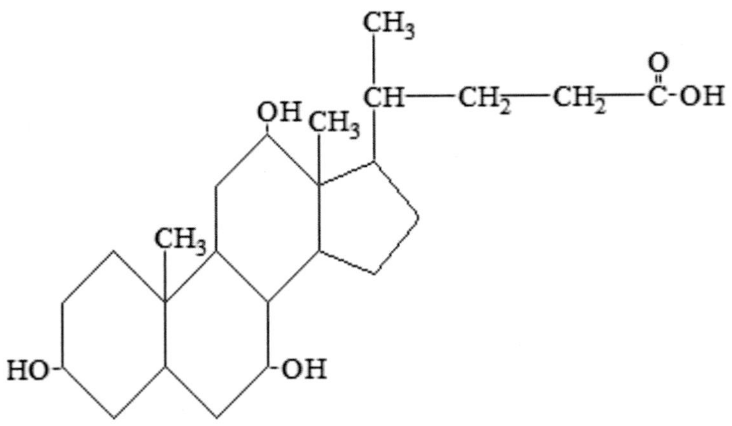

a. Use a grease pencil and mark a clean test tube (16 x 120 mm) with a B.

b. Use a graduated cylinder to add 5.0 mL of 1.0 %$_{m/v}$ bile salt solution to the test tube.

c. Use a transfer pipette and add 1 dropper full of corn oil to the test tube

d. Place a clean rubber stopper in the test tube and vigorously shake the test tube

e. Remove the stopper and set the test tube in a test tube rack.

4. Soap- The base hydrolysis of triglycerides produces soap. Soap is composed of salts of long chain fatty acids. The salt carboxylate end of the soap molecule gives it a degree of polarity while the long hydrocarbon chain is the nonpolar area. When enough of the hydrocarbon tails of soap dissolve into fat or grease, the polar salt carboxylate pull the fat/grease into solution with the water. The production of sodium palmitate from base hydrolysis of triglyceryl palmitate with sodium hydroxide is shown below.

Base Hydrolysis of Glyceryl Palmitate

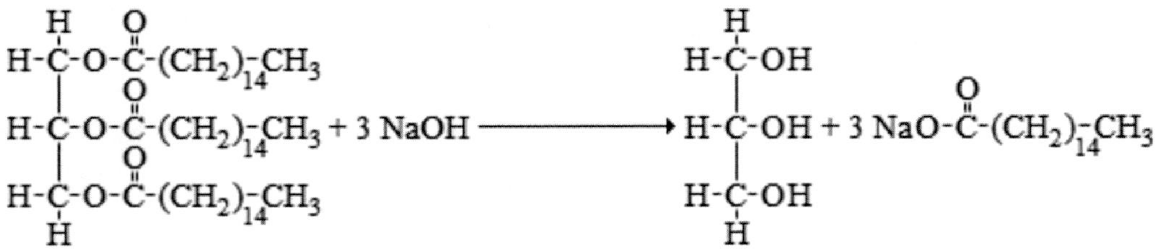

a. Use a grease pencil and mark a clean test tube (16 x 120 mm) with an S.

b. Use a graduated cylinder to add 5.0 mL of 1.0 %$_{v/v}$ bile soap solution to the test tube.

c. Use a transfer pipette and add 1 dropper full of corn oil to the test tube

d. Place a clean rubber stopper in the test tube and vigorously shake the test tube

e. Remove the stopper and set the test tube in a test tube rack.

5. Data Collection- In each of your test tubes there is an organic phase, in this case corn oil, above an aqueous phase. The relative effectiveness of each emulsifying agent can be evaluated by comparing the amount of the organic phase to the total amount of the organic/aqueous bilayer. The total distance from the bottom of a test tube to the top of the organic phase is an indication of the total amount of the bilayer. The distance from the separation of bilayer, between the aqueous phase and the organic phase, to the top of the organic phase indicates the amount of the organic phase.

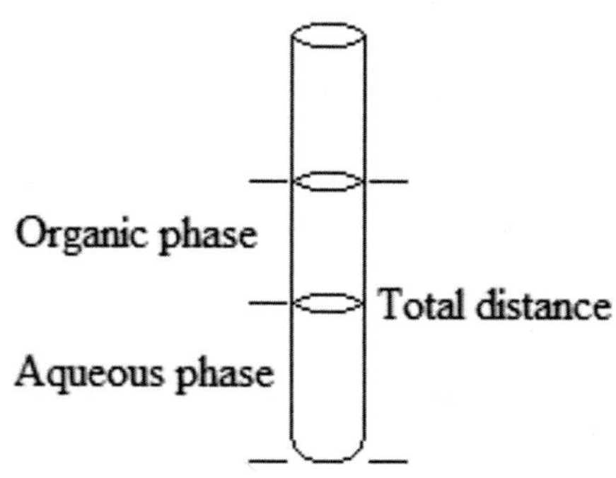

If an emulsifying agent completely brought the organic phase into solution with aqueous phase, the emulsifying agent would be considered one-hundred percent effective.

   a. Measure the total distance of your control test tube and your three experimental tubes to the closest 0.1 mm and record the results in the data table below.

   b. Measure the organic phase distance of your control test tube and your three experimental tubes to the closest 0.1 mm and record the results in the data table below.

   c. Divide the organic phase distance by the total distance and multiply the result by 100. Record the percentage in the data table below.

Emulsifying Agent Efficiency			
Mixture	**Organic phase, mm**	**Total distance, mm**	$\left(\dfrac{\text{Organic phase, mm}}{\text{Total distance, mm}}\right)100,\ \%$
Corn oil & Water			
Corn oil & 1.0% Ethanol			
Corn oil & 1.0% Bile Acid			
Corn oil & 1.0% Soap			

6. Analysis- As an emulsifying agent pulls more of the organic phase into the aqueous phase, the amount of the organic phase decreases and the efficiency percent increases.

a. Which emulsifying agent was most effective?

____

b. Which emulsifying agent was least effective?

____

7. Return your lab tray to its cabinet.

Notes:

Name: _______________________________

crn: _______________________________

Outcomes
Compare and contrast the primary, secondary, tertiary and quaternary protein structure. Define a peptide bond.

A. Prelab Assignment

Use printed or internet sources to draw the full structural formula of the amino acids below.

1. Tyrosine:

2. Lysine:

3. Aspartic acid:

4. Valine:

5. Asparagyl-aspatate:

B. Protein Structure

Proteins and peptides are polymers of amino acids in a specific sequence. There are 20 common amino acids found in humans. An amino acid consists of a central carbon with an amino group ($-NH_2$), a carboxylic acid group ($-COOH$), a hydrogen atom ($-H$), and a residue group ($-R$) attached. The simplest amino acid is glycine with its $-R$ being a hydrogen atom. Alanine's $-R$ is a methyl group ($-CH_3$).

$$\underset{\displaystyle R}{H\text{-}N\text{-}\overset{\displaystyle H}{\underset{|}{C}}\text{-}\overset{\displaystyle O}{C}\text{-}OH}$$

The amino acids are joined together by peptide bonds. Remember, the peptide bond is an amide bond between the amino end of one amino acid and the carboxylic acid end of another amino acid. Oligopeptides are small peptides, while larger peptides are polypeptides. A protein is a peptide with biological purpose and function. The sequence of the amino acids in a protein determine the shape and structure of the protein. The shape and structure of the protein determine its function. If the shape and structure is incorrect, the protein will lose functionality. Protein folding is the process of modifying and finishing a protein. There are four levels of structure in proteins, primary (1°), secondary (2°), tertiary (3°), and quaternary (4°).

Primary structure describes the sequence of amino acids and is read amino end (N-terminus) to the carboxylic acid end (C-terminus) of a protein. Both peptides and proteins both have primary structure because they both have a linear arrangement of amino acids formed into a chain through peptide bonding. If a protein loses its primary structure, it would be a solution of individual amino acids and be destroyed.

Secondary structure presents as alpha (α) helices and beta (β) strands, structures which are formed by hydrogen bonding between backbone N-H and C=O groups involved in the peptide bonds. To imagine an α helix picture a spiral of amino acids with the -R groups positioned to the outside of the spiral. A β strand can be described as a pleated or folded sheet with the –R to the outside of the sheet. Since the α helices and β strands are formed by the interaction of N-H and C=O groups, secondary structure is independent of the R-groups. Many physical (heat, agitation) and chemical (acids, bases, heavy metal ions, certain organic compounds) agents can interfere with secondary structure and alter the shape of the protein. When a protein's secondary structure is altered, it loses its functionality and is denatured. In some cases if the denaturing agent is removed, the protein can regain its biological function.

Interactions between their R-groups or between R-groups and backbone N-H and C=O groups, give proteins tertiary structure. Oxytocin is a simple nanopeptide (9 amino acids) where interaction between the fourth amino acid, cysteine, and the last amino acid, cysteine, form a disulfide bond between each other and cause the protein to fold back on itself. Many chemical and physical agents can interfere with tertiary structure and denature a protein. Again, if the denaturing agent is removed, the protein may, in some cases, be renatured.

Some proteins are combinations of two or more peptide chains that have to be brought together and folded into a functional shape. When this occurs the protein has quaternary structure. Insulin is an example of such a protein with a 21 amino acid α peptide chain and a 30 amino acid β chain. Hemoglobin is composed of two identical α chains of 141 amino acids each and two identical β chains of 146 amino acids each. Not every protein has quaternary structure, because not every protein is made up of two or more polypeptide chains. If the quaternary structure is compromised, the functionality of the protein is permanently lost since the chances of the constituent polypeptide chains being refolded into functional position is nil.

C. Denaturation

!!! SAFETY GOGGLES MUST BE WORN DURING ALL PROTEIN TESTING!!!

You will be testing several factors to see how they affect proteins. First describe the appearance of the contents in the control. Describe the color of the control contents and if the contents is clear or cloudy. For example, "white and cloudy" or "gray but clear" would be appropriate descriptions of the control. Next describe the contents of the experimental factor or factors. Record your results as less or more cloudy than your control. Cloudier is an incomplete statement. A complete statement would be, "Cloudier than control."

As a protein is denatured, it becomes less soluble in water; cloudier means the protein is less soluble than before you subjected it to the test factor. A protein can become so insoluble that it precipitates. This results in a clear liquid phase above the precipitate. Some test factors destroy the protein completely down to the individual amino acids. In this case, the solution will become clearer, but there will be no precipitate. Agitation acts on proteins by breaking the 4° and 3° level of structure. This causes the protein to spread out. Agitation also whips air into the protein and water, causing it to fluff up. Several questions will ask you to apply what you have learned today about denaturation to specific situations. In these cases, answer the question in terms of denaturation of proteins.

The photo below shows five test tubes with albumin. The first three tubes on the left have been exposed to a denaturing agent. The fourth tube from the left has been exposed to a substance but the albumin was not denatured. The fifth tube from the left was the control.

1. Heat

    a. Obtain a test tube rack and an electric heater from the supply bench. Fill a 250 ml beaker with tap water to the 150 ml mark. Place the beaker on the heater and turn it to high.

    b. Return to the supply bench with two clean test tubes (16 x 120 mm) and locate the $2.0\%_{m/v}$ albumin mixture. Swirl the albumin mixture to ensure that it is evenly distributed and then fill the test tubes two-thirds full with the albumin mixture.

    c. Take the two test tubes back to your lab space. Place one test tube in the test tube rack away from the heater and the other in the 250 ml beaker which is on the heater.

    d. Bring the water in the 250 ml beaker begins to boil. Let the experimental test tube boil for 5 minutes. Turn the heater off and allow the test tube to cool. When the test tube is cool to touch, remove it from the beaker and place it in the test tube rack.

e. Observe both test tubes and record their appearance in the data table below. When the electric heater has cooled return it to where you found it. Pour out and clean the experimental test tube but save the control test tube for your next experiment.

Test tube	Factor	Appearance
Experimental	**Heat**	
Control	**Room T^o**	

2. Heavy metal ions

a. Label your control test tube with a C. Label four clean test tubes (16 x 120 mm) 1, 2, 3, & 4 and place them in the test tube rack with your control.

b. Take the test tube rack with test tubes to the supply bench. Remember to swirl the albumin mixture to ensure that it is evenly distributed and then add about 4-5 ml (2 fingers) of 2.0%$_{m/v}$ albumin mixture to test tubes 1 through 4.

c. Add 3-4 drops of $HgCl_2$ to test tube 1, 3-4 drops of $AgNO_3$ to test tube 2, 3-4 drops of $Pb(C_2H_3O_2)_2$ to test tube 3, 3-4 drops of $NaCl$ to test tube 4, and 3-4 drops of H_2O to the control test tube.

d. Gently swirl all the test tubes and then let them set for 2-4 minutes.

e. Observe all test tubes and record their appearance in the data table below. Dispose of the $HgCl_2$ albumin mixture in the $HgCl_2$ waste bottle. Dispose of the contents the remaining experimental test tubes in the waste jug. Rinse and clean your experimental test tubes and save the control test tube for your next experiment.

Test tube	Factor	Appearance
Exp. 1 $HgCl_2$	Hg^{+2}	
Exp. 2 $AgNO_3$	Ag^{+}	
Exp. 3 $Pb(C_2H_3O_2)_2$	Pb^{+2}	
Exp. 4 $NaCl$	Na^{+}	
Control	H_2O	

3. Organic compounds

a. Label four clean test tubes (16 x 120 mm) 1, 2, 3, & 4 and place them in the test tube rack with your control.

b. Take the test tube rack with test tubes to the supply bench. Remember to swirl the albumin mixture to ensure that it is evenly distributed and then add about 4-5 ml (2 fingers) of $2.0\%_{m/v}$ albumin mixture to test tubes 1 through 4.

c. Add 2 ml of ethanol to test tube 1, 2 ml of acetone to test tube 2, 2 ml of caffeine solution to test tube 3, and 2 ml of tannic acid solution to test tube 4.

d. Gently swirl all the test tubes and then let them set for 2-4 minutes.

e. Observe all test tubes and record their appearance in the data table below. Dispose of the experimental albumin mixtures by pouring down the drain of your lab sink. Rinse and clean your experimental test tubes and save the control test tube for your next experiment.

Test tube	Factor	Appearance
Exp. 1	Ethanol	
Exp. 2	Acetone	
Exp. 3	Caffeine	
Exp. 4	Tannic Acid	
Control	H_2O	

4. Acids and bases

a. Label two clean test tubes (16 x 120 mm) 1 & 2 and place them in the test tube rack with your control.

b. Take the test tube rack with test tubes to the supply bench. Remember to swirl the albumin mixture to ensure that it is evenly distributed and then add about 4-5 ml (2 fingers) of $2.0\%_{m/v}$ albumin mixture to test tubes 1 and 2.

c. Add 10 drops of 1 M HCl to test tube 1 and 10 drops of 1 M NaOH to test tube 2.

d. Gently swirl all the test tubes and then let them set for 2-4 minutes.

e. Observe all test tubes and record their appearance in the data table below. Dispose of the experimental albumin mixtures by pouring down the drain of your lab sink. Rinse and clean your experimental test tubes and save the control test tube for your next experiment.

Test tube	Factor	Appearance
Exp. 1	HCl	
Exp. 2	NaOH	
Control	H_2O	

5. Agitation

a. Label one clean test tubes (16 x 120 mm) E for experimental and place it in the test tube rack with your control.

b. Take the test tube rack with test tubes to the supply bench. Remember to swirl the albumin mixture to ensure that it is evenly distributed and then add about 4-5 ml (2 fingers) of $2.0\%_{m/v}$ albumin mixture to test tube E along with a paper clip. Remember to swirl the albumin mixture to ensure that it is evenly distributed. Seal both your test tubes with a rubber stopper.

c. Return to your lab space with your test tube rack and test tubes. Vigorously shake test tube E for 2-4 minutes.

d. Observe both test tubes and record their appearance in the data table below. Rinse and clean your all test tubes. Return your lab tray to its cabinet.

Test tube	Factor	Appearance
Experimental	Agitated	
Control	Not Agitated	

6. Analysis

    a. List the two physical agents and four chemical agents that denature proteins.

____

____

    b. Name the ions that denatured the albumin?

____

____

    c. Name the organic compounds that denatured the albumin?

____

____

    d. What effect did changing pH have on the albumin? Remember the pH was lowered using the acid and was raised using the base.

____

____

    e. What levels of protein structure are affected when proteins are denatured?

____

____

    f. Can denatured proteins be changed back to their functional state?

____

____

Notes:

Name: ______________________________________

crn: ______________________________________

Outcomes
Compare and contrast the structure of DNA and RNA Describe the processes of DNA replication, transcription and translation Use the genetic code to determine an amino acid sequence from a sequence of DNA nucleotides.

A. Prelab Assignment

Use printed or internet sources to draw the relaxed ladder model of DNA and RNA.

1. Deoxyribonucleic Acid (DNA):

2. Ribonucleic Acid (RNA):

3. Deoxyribonucleic acid is a long helical polymer of nucleotides. DNA is the genetic code of the chromosomes. The genetic code determines the sequence of amino acids in a peptide or protein and the sequence of the amino acids determine the shape which determines the function of the protein. DNA is replicated during the S phase of interphase prior to cell division. During replication the DNA unwinds and unzips. An enzyme, DNA polymerase, assembles and aligns free nucleotides using complementary base pairing. The nucleotide cytosine is always paired with guanine and adenine is always paired with thymine during replication. The nucleotides are held together by hydrogen bonding. Study the diagram of replication and then fill in the missing nitrogen bases. Let A = adenine, T = thymine, G = guanine, and C = cytosine.

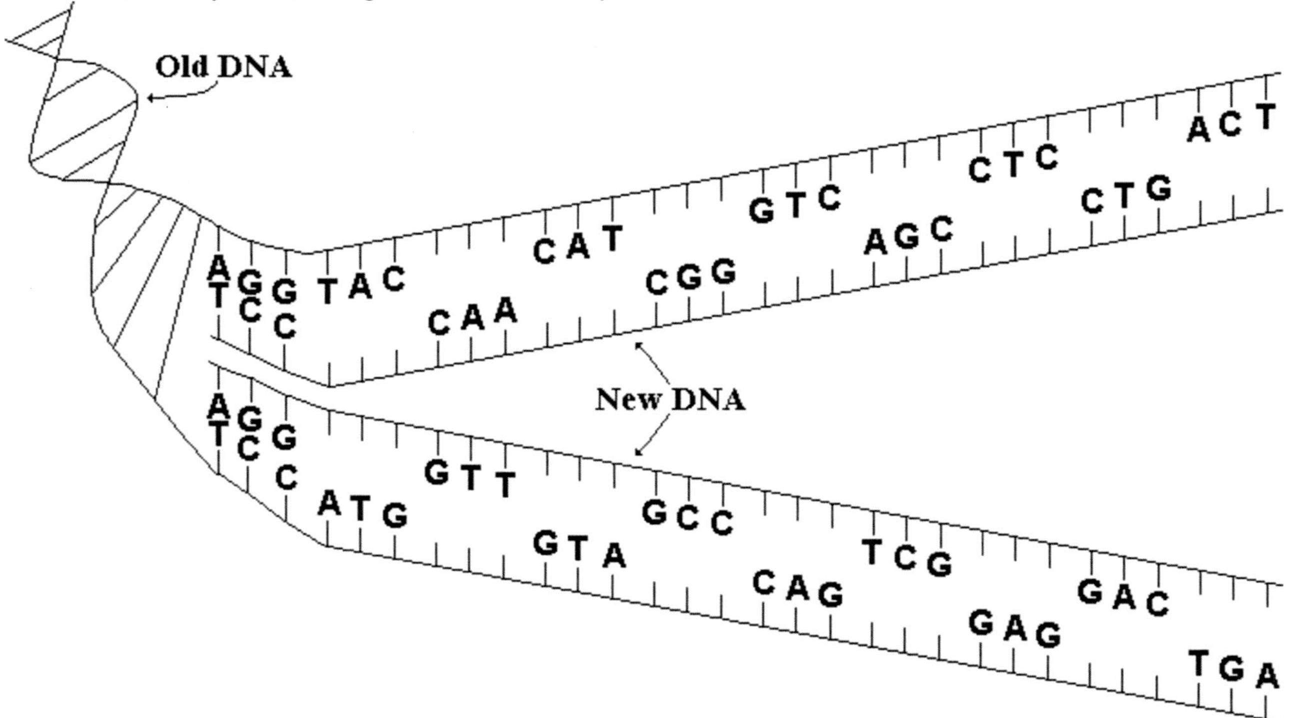

4. Protein synthesis begins when the genetic code for the amino acid sequence of a protein is transcribed to mRNA. The mRNA then takes the code out to cytoplasm and the code is read by the ribosomes to produce a protein. Transcription begins when a portion of the DNA unwinds and unzips. RNA polymerase is an enzyme present in the nucleus that assembles free RNA nucleotides. The RNA polymerase, next aligns the free RNA nucleotides with complementary DNA nucleotides in the unzipped portion of template DNA segment using complementary base pairing. Complementary base pairing for DNA and RNA is almost the same as complementary base pairing for DNA alone when it replicates. The base pair complements for DNA and RNA are cytosine always pairs guanine, guanine always pairs cytosine. Thymine in the DNA template segment results in adenine in the mRNA strand. However, since uracil replaces thymine in RNA, adenine in the DNA template segment results in uracil in the mRNA. When the mRNA segment is completed, it separates from the DNA and moves out into the cytoplasm to complete protein synthesis. In the transcription diagram below, fill in the missing nitrogen bases in the DNA coding and template segments and mRNA strand.

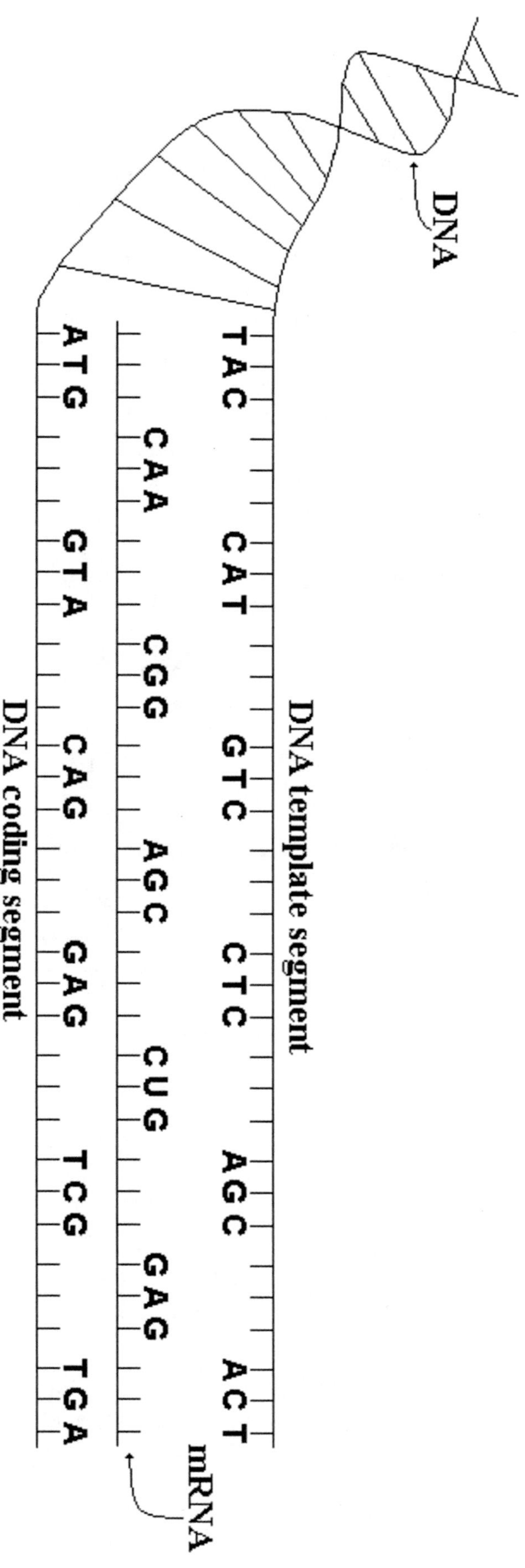

DNA
DNA template segment
TAC CAT GTC CTC AGC ACT
mRNA
CAA CGG AGC CUG GAG GAG
DNA coding segment
ATG GTA CAG GAG TCG TGA

B. Translation

Translation is when the genetic code of nucleic acids is translated into a protein for the cell to use in its metabolism or structure.   There are four steps in translation- activation, initiation, elongation, and termination.   Activation begins when the tRNA segments bond to their specific amino acids.   The bonding of the amino acid to tRNA activates its anticodon. At the beginning of initiation, the start codon, AUG, of mRNA bonds to a small subunit of a ribosome. Next, the tRNA anticodon with methionine base pairs with mRNA start codon and the large subunit of the ribosome joins the small subunit to initiate protein synthesis. The start of elongation is when the complementary tRNA anticodon with its specific amino acid attaches to the next mRNA codon after the start codon. The new amino acid, then forms a peptide bond with methionine and the tRNA is released from the methionine. Several ribosomes can translate same strand of mRNA and translation continues until the stop codon is reached. Then the protein is released from the ribosome and methionine is removed from peptide chain. Although all proteins have methionine at the start of the protein chain, it is removed in the finished protein. On the next page is a diagram of the overall process of protein synthesis. The small subunit has been left out of the diagram for the sake of simplicity.

The table below lists all the mRNA codons and the amino acids that they code for. Notice that some codons code for the same amino acid although they have different base sequences. There are 64 mRNA codons and only 20 amino acids. Each mRNA codon has a complementary tRNA anticodon, except the three mRNA stop codons. Stop codons terminate translation and do not have tRNA anticodons. If you know the mRNA codon, you also know its DNA triplet and its tRNA anticodon. For example, if the mRNA codon is AGU, then its DNA triplet is TCA and its tRNA anticodon is UCA. However, given the mRNA codon is UGA, its DNA triplet is ACT but it has no tRNA anticodon. Study the table below and then proceed to **1.** below.

The Genetic Code Table – mRNA codons

UUU phenylalanine	UCU serine	UAU tyrosine	UGU cystine
UUC phenylalanine	UCC serine	UAC tyrosine	UGC cystine
UUA leucine	UCA serine	UAA stop	UGA stop
UUG leucine	UCG serine	UAG stop	UGG tryptophan

CUU leucine	CCU proline	CAU histidine	CGU arginine
CUC leucine	CCC proline	CAC histidine	CGC arginine
CUA leucine	CCA proline	CAA glutamine	CGA arginine
CUG leucine	CCG proline	CAG glutamine	CGG arginine

AUU isoleucine	ACU threonine	AAU asparagine	AGU serine
AUC isoleucine	ACC threonine	AAC asparagine	AGC serine
AUA isoleucine	ACA threonine	AAA lysine	AGA arginine
AUG methionine, start	ACG threonine	AAG lysine	AGG arginine

GUU valine	GCU alanine	GAU aspartic acid	GGU glycine
GUC valine	GCC alanine	GAC aspartic acid	GGC glycine
GUA valine	GCA alanine	GAA glutamic acid	GGA glycine
GUG valine	GCG alanine	GAG glutamic acid	GGG glycine

G C
A T
DNA
TAC CGT ACG TGC TTT ACG
AUG GCA UGC AAA ACG TGA
ATG GCA TGC AAA ACG TGA
AUG GCA UGC UUU ACG ACU
Nuclear membrane
mRNA
Ala
Cys
Phe
Thr
Tetrapeptide
(a small protein)
Met
Ala
Cys
Phe
Ribosome
AUG GCA UGC UUU ACG UGA
AAA
UGC
Start codon
AUG GCA UGC UUU ACG UGA
codons
Stop codon
Amino Acids
Met
UAC
Phe
AAA
Thr
UGC
anticodon
Cys
ACG
Ala
CGU
tRNA

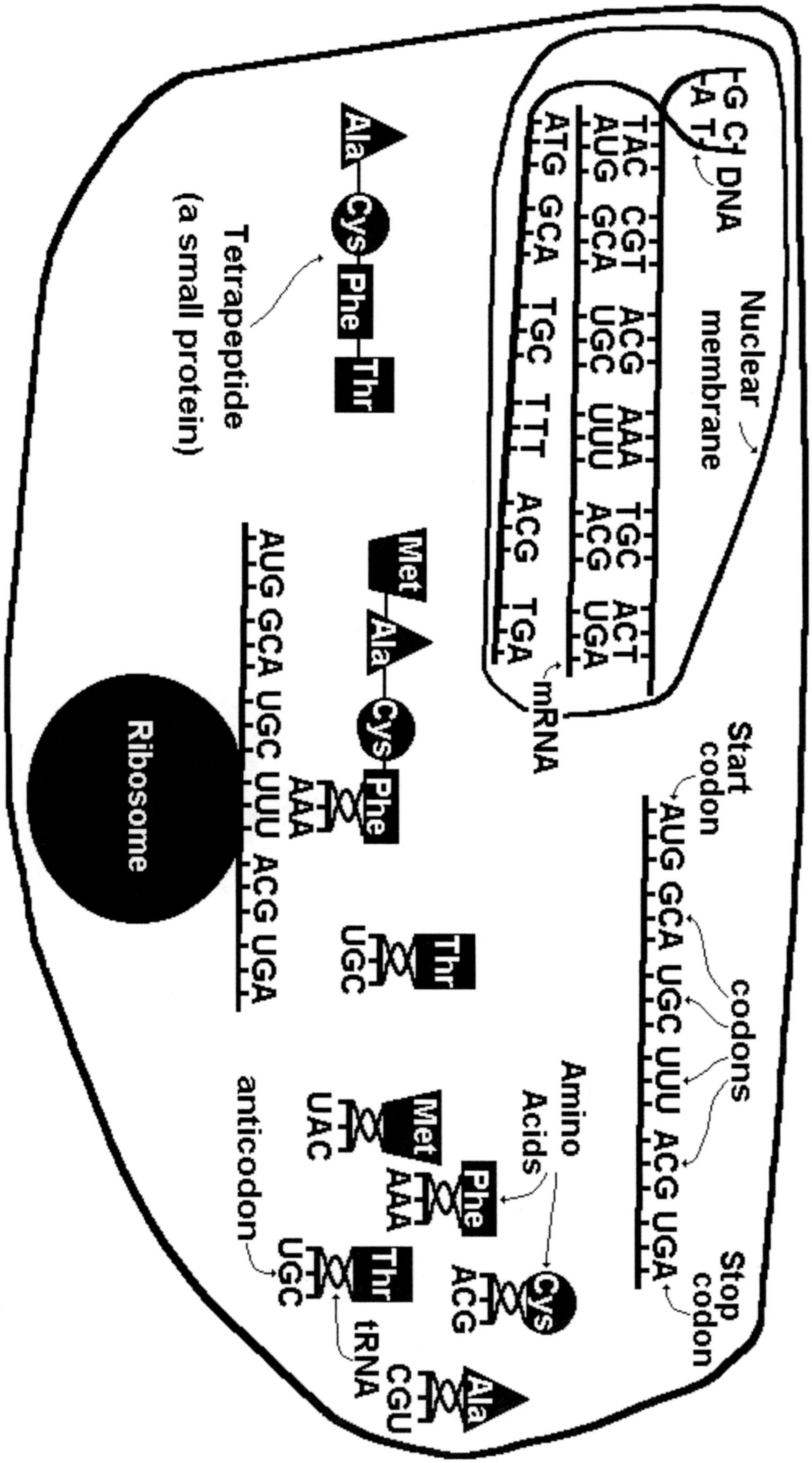

1. In the diagram shown below, fill in the missing bases for the tRNA anticodons and the DNA triplets, then use **The Genetic Code Table** determine the amino acids and/or actions being coded for. If a mRNA codon has no tRNA anticodon, write the word none in the tRNA anticodon space.

DNA template					
mRNA codons	AUG	UCU	GAA	GCU	UAG
tRNA anticodons					
Amino Acids					

2. In the diagram shown below, fill in the missing bases for the tRNA anticodons and the DNA triplets, then use **The Genetic Code Table** determine the amino acids and/or actions being coded for. If a mRNA codon has no tRNA anticodon, write the word none in the tRNA anticodon space.

DNA template					
mRNA codons	AUG	UAC	CGA	GUC	UAA
tRNA anticodons					
Amino Acids					

3. Thyroxin Releasing Factor (TRF) is produced in the anterior pituitary gland and regulates, as its name implies, the release of thyroxin from the thyroid. TRF is a simple tripeptide with the following amino acid sequence: glutamic acid-histidine-proline. How many different DNA sense sequences will code for the production of TRF? Use the **The Genetic Code Table** to determine the number of DNA sense sequences that will code for the synthesis of TRF. Remember that every DNA sense strand must have a mRNA coding strand and that every mRNA coding strand must have a start and a stop codon.

    a. Write the mRNA codon that always starts a mRNA coding sequence.

____

b. Write all the mRNA codons that code for the amino acid glutamic acid.

____

c. Write all the mRNA codons that code for the amino acid histidine.

____

d. Write all the mRNA codons that code for the amino acid proline.

____

e. Write all the mRNA codons that code for the stop of mRNA coding sequence.

____

f. Use the information above to calculate the total number of mRNA coding sequences that will cause the synthesis of TRF. Show all calculations and underline your answer to receive full credit.

g. What would be the selective advantage of having a genetic code that has several DNA sense sequences that all result in the same protein?

____

____

____

____

____

C. DNA Isolation

!!! SAFETY GOGGLES MUST BE WORN DURING DNA ISOLATION!!!

You will now extract DNA from your inner cheek cells in your mouth. If you follow the directions, at the end of this lab period you will have a sample of your own DNA.

1. Secure a screw-cap culture tube and test tube rack to hold the culture tube from the lab bench. Remove the cap from the tube and set it aside. Add distilled/deionized water to the culture tube until it is about one third full. Place the cap loosely on the culture tube and set the tube in the test tube rack.

2. Secure a screw-cap culture tube and test tube rack to hold the culture tube from the lab bench. Remove the cap from the tube and set it aside. Add distilled/deionized water to the culture tube until it is about one third full. Place the cap loosely on the culture tube and set the tube in the test tube rack.

3. Add 3 drops of plasma membrane lysis liquid to your culture tube. Place the screw-cap securely on the culture tube and gently invert the tube five times to rupture the cell and nuclear membranes.

4. Remove the cap from the culture tube and use a disposable transfer pipette to add ten drops of 5.0M NaCl solution to the culture tube. Replace the screw-cap securely on the culture tube and gently invert the tube five times.

5. Use a grease pencil to mark your initials on the culture tube. Place your culture tube in a hot water bath at 55°C for ten minutes to digest all cell proteins.

6. Remove the collection tube from the water bath at the end of ten minutes. Carefully open the tube. Hold the tube at a 45° angle and gently add ice-cold ethanol until the tube is almost full. Replace the cap and gently place the test tube in your test tube rack. Observe the culture tube at the interface of the ice-cold ethanol and DNA solution. You should see your DNA beginning to precipitate.

7. At the end of five minutes, make sure the cap or stopper is firmly in place and very gently invert the test tube. This will complete the precipitation of your DNA. Place the culture tube in a rack and let your DNA settle to the bottom of the tube.

8. When your DNA has settled to the bottom, show it to your instructor. If you have not been successful at isolating DNA, you may wish to repeat the procedure. Part of your grade for this lab is based on your ability to successfully isolate DNA.

9. After you have shown your DNA to your instructor, label it and turn it in to your instructor. A complete label will include- *Homo sapiens* DNA in ethanol, the date, and your initials. Return your lab tray to its cabinet.

10. In the space below briefly describe the appearance of your DNA.

____

____

____

____

____